Palgrave Studies in Natural Resource Management

Series editor
Justin Taberham
London, UK

More information about this series at
http://www.palgrave.com/gp/series/15182

Robert C. Brears

Natural Resource Management and the Circular Economy

palgrave
macmillan

Robert C. Brears
Mitidaption
Christchurch, Canterbury, New Zealand

Palgrave Studies in Natural Resource Management
ISBN 978-3-319-71887-3   ISBN 978-3-319-71888-0   (eBook)
https://doi.org/10.1007/978-3-319-71888-0

Library of Congress Control Number: 2017963063

Cover illustration: Getty/Kelly Sillaste
Cover Design: Fatima Jamadar

Printed on acid-free paper

This Palgrave Macmillan imprint is published by Springer Nature
The registered company is Springer International Publishing AG
The registered company address is: Gewerbestrasse 11, 6330 Cham, Switzerland

Series Editor Foreword

Series Foreword

Natural Resource Management (NRM)

The World Bank definition of NRM is as follows:

'The sustainable utilization of major natural resources, such as land, water, air, minerals, forests, fisheries, and wild flora and fauna. Together, these resources provide the ecosystem services that underpin human life.'

Natural Resource Management covers a very wide range of interwoven resource areas, management processes, threats, and constraints, including aquatic ecosystems, natural resources planning, and climate change impacts. Similarly, NRM professionals are very diverse in their qualifications and disciplines.

There is a significant and growing sector for NRM services and the worldwide market for this sector was almost $30 billion in 2015, according to *Environment Analyst*.

This book series will focus on applied, interdisciplinary, and cross-sectoral approaches, bringing together professionals to publish titles across the global sector.

The series will focus on the management aspects of NRM and titles will cover

- Global approaches and principles
- Threats and constraints
- Good (and less good) practice
- Diverse and informative case study material from practitioners and applied managers
- Cutting-edge work in the discipline

The issues covered in this series are of critical interest to advanced level undergraduates and master's students as well as industry, investors, and practitioners.

CEnv   Justin Taberham
Series Editor
www.justintaberham.com

Acknowledgements

I wish to first thank Rachael Ballard who is not only a wonderful commissioning editor but a visionary who enables books like mine to come to fruition. I wish to also thank mum who has a great interest in the environment and natural resource-related issues and has supported me in this journey of writing the book.

Introduction

Since the Industrial Revolution, the total amount of waste has constantly grown as economic growth has been based on a 'take-make-consume-dispose' model. This linear model assumes resources are abundant, available, and cheap to dispose of. While the current linear economic model has generated an unprecedented level of growth, it has led to constraints on the availability of natural resources due to rising demand, generation of waste, and environmental degradation.

From the sustainable development perspective, the linear economy is leading to the rapid accumulation of human and physical capital at the expense of natural capital, impacting the ability of current generations to ensure future generations have at least the same level of welfare. While weak sustainability proponents argue that depleted natural capital can be replaced by even more valuable physical and human capital, the strong view is that natural capital should be protected, not depleted, due to it being exhaustible, often unevenly distributed geographically, limited in availability at times, and undervalued, as associated benefits, including their non-use benefits, are not reflected in market prices of natural resources.

Around the world, there is a move towards a 'circular economy' where products and waste materials are reused, repaired, refurbished, and recycled with significant economic and environmental benefits. A key aspect of the circular economy is that materials, which have accumulated in the economy, constitute important man-made stocks that can be exploited through recycling to gain secondary raw materials and reused and remanufactured to keep products in the commercial lifecycle. The overall aim of the circular economy is to decouple economic growth from resource use and associated environmental impacts.

Government intervention has an important role in developing the circular economy and encouraging a life cycle perspective to be taken by economic actors. In particular, governments can use a variety of innovative policy tools, both fiscal and non-fiscal in nature, including environmental taxes and charges, subsidies and incentives, and tradeable permits as well as regulations, business support mechanisms, and information and awareness campaigns, to encourage businesses to design out waste throughout the value chain rather than rely on solutions at the end of a product's life, facilitate access to financial capital for businesses developing circular economy innovations, provide research and development funding for circular economy technologies, support entrepreneurs and small-to-medium enterprises developing new circular economy markets, and facilitate better consumption choices by consumers.

This book contains case studies on how London, Seattle, Flanders, New South Wales, Denmark, Germany, the Netherlands, and Scotland—considered global leaders in the development of the circular economy—use fiscal and non-fiscal tools to develop the circular economy and encourage a life cycle perspective to be taken by economic actors in an attempt to decouple economic growth from resource use and associated environmental impacts.

The case studies are chosen for the following reasons. London is a global economic leader that aspires to promote circular economy innovations; Seattle is a leading industrial and information technology hub that is known for developing progressive technologies; Flanders is a highly productive region of Belgium and is recognised internationally as a circular economy hub; New South Wales is Australia's largest state in terms of economic size and is advancing the development of clean technologies and other resource-efficient practices; Denmark has a long history of sustainable development initiatives and therefore is considered a leader in circular economy technologies; Germany has long been recognised as a trailblazer in environmental technologies and practices; the Netherlands, while small in terms of land size, has a high level of productivity that is being harnessed in the development of a circular economy; and Scotland is internationally recognised as a leading example of how a fossil fuel dominated economy can transition towards a circular, low-carbon economy.

Following the case studies, a series of best practices have been identified for other locations around the world implementing a variety of fiscal and non-fiscal tools to develop the circular economy and encourage a life cycle perspective to be taken by economic actors in an attempt to decouple economic growth from resource use and associated environmental impacts.

The book's chapter synopsis is as follows:

Chapter 1 introduces readers to the numerous challenges faced by the current linear economic model followed by an overview of the circular economy.

Chapter 2 provides readers with a review of the various fiscal and non-fiscal tools available to develop the circular economy along with mini-case examples of policies being implemented around the world.

Chapters 3, 4, 5, 6, 7, 8, 9, and 10 provides readers with case studies on the development of the circular economy in London, Seattle, Flanders, New South Wales, Denmark, Germany, the Netherlands, and Scotland.

Chapter 11 provides readers with a series of best practices for other locations around the world developing a circular economy.

Chapter 12 concludes from the case studies that developing a circular economy is not a static activity.

Contents

List of Tables

1

The Circular Economy

Introduction

Our current industrialised economy is essentially a linear model in which resource consumption follows a 'take-make-consume-dispose' pattern where natural resources are harvested for the manufacturing of products, which are then disposed of after consumption. In terms of volume, around 65 billion tonnes of raw materials entered the economic system in 2010 and this figure is expected to increase to around 82 billion tonnes in 2020.[1] It has become increasingly clear that this economic model is untenable due to a growing shortage of materials, increased levels of pollution, increased material demand, and a growing demand for responsible products by consumers. In contrast, the circular economy aims to reduce resource consumption, recover materials, and recycle waste into new products and materials with the aim of decoupling economic growth from resource use and associated environmental impacts.

© The Author(s) 2018
R. C. Brears, *Natural Resource Management and the Circular Economy*,
Palgrave Studies in Natural Resource Management,
https://doi.org/10.1007/978-3-319-71888-0_1

The Linear Economy

In our current economic model, manufactured capital, human capital, and natural capital all contribute to human welfare by supporting the production of goods and services in the economic process, where natural capital—the world's stock of natural resources (provided by nature before their extraction or processing by humans)—is typically used for material and energy inputs into production and acts as a 'sink' for waste from the economic process.[2] This economic model can be best described as 'linear' which typically involves economic actors[3]—who are people or organisations engaged in any of the four economic activities of production, distribution, consumption, and resource maintenance—harvesting and extracting natural resources, using them to manufacture a product, and selling a product to other economic actors, who then discard it when it no longer serves its purpose.[4] As such, natural resources in the linear economic model:

- *Become inputs*: Material resources used in the economy come from raw materials that are extracted from domestic natural resource stocks or extracted from natural resource stocks abroad and imported in the form of raw materials, semi-finished materials, or materials embedded in manufactured goods. Material resources are extracted with the usable parts of the resources entering the economy as material inputs where they become priced goods that are traded, processed, and used. Other parts remain unused in the environment and are called unused materials or unused extraction.
- *Become outputs*: After use in production and consumption activities, materials leave the economy as an output either to the environment in the form of residuals (pollution, waste) or in the form of raw materials, semi-finished materials, and materials embedded in manufactured goods.
- *Accumulate in man-made stocks*: Some materials accumulate in the economy where they are stored in the form of buildings, transport infrastructure, or durable and semi-durable goods such as cars, industrial machinery, and household appliances. These materials are eventually released in the form of demolition waste, end-of-life vehi-

cles, e-waste, bulky household waste, and so on, which if not recovered flow back to the environment.

- *Create indirect flows*: When materials or goods are imported for use in an economy, their upstream production is associated with unused materials that remain abroad including raw materials needed to produce the goods and the generation of residuals. These indirect flows of materials consider the life cycle dimension of the production chain but are not physically imported. As such, the environmental consequences occur in countries from which the imports originate[5]

Linear Economy Challenges

While the current linear economic model has generated an unprecedented level of growth, the model has led to constraints on the availability of natural resources due to rising demand, in addition to the generation of waste and environmental degradation from a variety of challenges.

Economic Growth

The world economy is projected to grow on average by just over 3 percent per annum over the period 2014–2050, resulting in the global economy doubling in size by 2037 and nearly tripling by 2050. During this time, there will be a shift in economic power away from the established advanced economies in North America, Europe, and Japan towards the emerging economies. For example, rapid economic growth in Mexico and Indonesia could result in these countries having larger economies than the UK and France by 2030, in purchasing power parity terms, while other economies including Nigeria and Vietnam could grow at 5 percent or more per annum by 2050 compared to projected growth of 1.5–2.5 percent in advanced economies.[6] To date, rising global consumption and the industrialisation of developing countries, in addition to globalisation, has led to the rapid increase in global trade with the physical trade volume more than doubling in total between 1980 and 2010. While there was a slowdown in trade in the early 1980s due to the second oil crisis and in 2009 from the global financial crisis, the physical trade volume between these periods increased on average by 2.4 percent per annum.[7] It is estimated that if the global economy carries on in a business-as-usual manner regarding resource consumption, we will need by 2050, on aggregate, the equivalent of two planets to sustain us.[8]

Changing Consumption Patterns

As income levels rise, changes in spending patterns occur when individuals move from a very low income (annual wages of less than USD $1000 per annum) to a lower middle income (between USD $3000 and USD $5000). For instance, as income levels rise, individual spending on food falls from more than 40 percent of the total income to around 10 percent. Regarding expenditure on types of foods, there is a strong positive correlation between the level of income and the consumption of animal protein with the consumption of meat, milk, and eggs increasing at the expense of staple foods.[9] Meanwhile, demand for clothing increases significantly as income rises, with emerging markets projected to make up 57 percent of total global clothing demand in 2050, up from 35 percent in 2012. Housing and furniture demand will increase with rising income levels too: In China, refrigerator ownership per 100 rural households increased from 0.1 to 17.8 between 1980 and 2004.[10] Demand for smartphones, tablets, and LCD/LED TVs will result in revenues for the consumer electronics market reaching nearly USD $3 trillion in 2020, up from around USD $1.45 trillion in 2015, with rising disposable income among the middle class in China and India expected to contribute significantly to this demand.[11]

Raw Material Scarcity

With rising income levels and economic growth around the world, raw material extraction is increasing to meet demand for both high-tech products and everyday consumer products including mobile phones, synthetic fuels, lithium-ion batteries, thin-layer photovoltaics, and so on. It is estimated that annual global material extraction will reach 183 billion tonnes by 2050, which is more than double the amount in 2015.[12]

Because of rising demand, global extraction of minerals is increasing while the ore grades being mined are declining; for instance, ore grades in Australia have declined by a factor of 2–5 since the beginning of mining in the country. Similarly, the grade of iron ore in the United States over the past century has declined from 60 percent to 20 percent.[13] There is also likely to be shortages of critical raw materials with the European Commission forecasting the shortage of 14 out of 41 minerals and metals analysed. Their high supply risk is mainly due a high share of the worldwide production coming from only a handful of countries including China (magnesium, rare earths, tungsten, etc.), Russia (platinum group metals), the Democratic Republic of Congo (cobalt, tantalum), and Brazil (niobium and tantalum), with many of these countries using trade, taxation, and investment instruments to reserve their resource base for their exclusive use.[14]

Volatility of Resource Prices

During most of the twentieth century, resource prices, including food, energy, and steel, declined despite rising populations and economic growth because of new low-cost sources of supply and technological innovation. However, in the first decade of the twenty-first century alone, price rises have varied significantly; for example, energy prices have increased by 190 percent, food prices by 135 percent, and material prices by 135 percent. The volatility of food, agricultural raw materials, and metal prices has also increased over the period 2000–2010, with the average annual volatility of resource prices being more than three times that over the course of the twentieth century and more than 50 percent higher than in the 1980s. Over the next quarter century, there will likely be resource-related shortages due to a variety of factors including an increase in the number of middle-class consumers, increased demand for new sources of supply, and extraction becoming more challenging and expensive.[15] High and volatile resource prices can present serious economic and social challenges by restricting market access, hampering investment, and even undermining peace and security; for example, future swings in food prices will jeopardise food security in low-income countries.[16,17]

Population Growth

In 2017, the world's population reached 7.6 billion with the world adding 1 billion inhabitants over the past 12 years. Sixty percent of the world's people live in Asia, 17 percent in Africa, 10 percent in Europe, 9 percent in Latin America and the Caribbean, and the remainder in North America and Oceania. The world's population is projected to reach 8.6 billion in 2040 and increase to 9.8 billion in 2050 and 11.2 billion in 2100. Between now and 2050, more than half of this growth will occur in Africa: of the additional 2.2 billion people who may be added between 2017 and 2050, 1.3 billion will be added in Africa. Globally, much of the overall increase in population between now and 2050 will occur in either high-fertile countries or in countries with large populations. Between now and 2050, it is estimated that half of the world's population growth will occur in just nine countries: India, Nigeria, Democratic Republic of Congo, Pakistan, Ethiopia, Tanzania, the United States of America, Uganda, and Indonesia (in order of their expected contribution to total growth).[18]

Rapid Urbanisation

Currently, 54 percent of the world's population lives in urban areas and by 2050 this is expected to increase to 66 percent. Projections show that overall growth of the world's population and rate of urbanisation could add an additional 2.5 billion people to urban populations by 2050, with 90 percent of this increase in Asia and Africa. In 2014, the world's urban population was 3.9 billion and this is expected to surpass 6 billion by 2045. The number of 'mega-cities' with 10 million or more inhabitants has risen from 10 in 1990 to 28 in 2014. By 2030, the world is projected to have 41 mega-cities. Meanwhile, the fastest-growing settlements are medium-sized cities and cities with less than 1 million inhabitants in Asia and Africa. Between 2000 and 2014, the world's cities with more than 500,000 inhabitants grew on average 2.4 percent per annum, but 43 of these cities grew twice as fast, with average growth rates of over 6 percent per annum.[19] Some of the impacts of urbanisation include the reduction of vegetation, the alteration of surface water flows, the alteration of surface energy with the build-up of the urban heat island effect, increased energy consumption for electricity, transportation, and heating, and an increase in waste.[20,21] For example, in Shanghai, urbanisation has led to water supply increasing from 180 to 3090 million cubic metres between 1949 and 2010, with tap water sold for industrial use increasing from 31 to 580 million cubic metres over the same period. Similarly, electricity consumption in the city increased over the period 1950–2010 at a rate of 16,171 million kilowatt hours per decade with industrial consumption rising from 756 to 78,661 million kilowatt hours.[22]

Rising Infrastructure Demand

With resource scarcity increasing due to population growth, both the public and private sector are expanding into evermore challenging extraction environments that have higher costs and are more reliant on new technologies. For instance, to meet future demand for steel, water, agricultural products, and energy it would require a total investment of around $3 trillion per annum, which is $1 trillion more than spent in recent history and does not include measures to help populations adapt to the effects of climate change.[23] In addition, climate change is continuing to drive investments in water resources, renewable energy, and clean technologies;[24] for example, one study found that the costs of adaptation are 1–2 percent of baseline costs (assumption of no climate change) for all Organisation for Economic Co-operation and Development (OECD) countries with the main element being the extra costs of developing water resources to meet higher levels of municipal water demand.[25]

Energy Use

It is projected that global energy consumption will increase by 48 percent between 2012 and 2040. While renewables and nuclear power will be the fastest-growing sources of non-fossil fuel energy, increasing on average by 2.6 percent per year and 2.3 percent per year through to 2040 respectively, fossil fuels will continue to provide most of the world's energy with liquid fuels, natural gas, and coal accounting for 78 percent of total world energy consumption in 2040. Meanwhile, electricity consumption by end users will rise by 1.9 percent/year on average from 2012 to 2040 with the largest growth in non-OECD countries at 2.5 percent/year, as rising living standards increase demand for home appliances and electronic devices as well as for public administration and commercial services such as schools, office buildings, and retail stores.[26,27] In Vietnam, for example, it is estimated that as the economy grows between 2016 and 2030, the percentage of households owning a refrigerator will increase from 63.7 percent to 94.9 percent.[28]

Water Degradation

By 2030, global demand for water will outstrip supply by 40 percent, and by 55 percent in 2050. This increase in demand will come mainly from manufacturing (+400 percent), electricity (+140 percent), and domestic use (+130 percent). By 2050, it is estimated that 3.9 billion people—40 percent of the world's population—will be living in river basins under severe water stress. Already in many areas of the world, groundwater is being exploited faster than it can be replenished leading to land subsidence and saltwater intrusion. Globally, over 80 percent of the world's wastewater and over 95 percent in some least developed countries is released to the environment without treatment, resulting in polluted rivers, lakes, and coastal waters.[29,30,31] Water quality is projected to deteriorate further through nutrient flows from agriculture: In the United States, farmland occupies less than a third of the land but US Environmental Protection Agency has deemed that agricultural activities impair more US streams than any other class of human impacts (around 40 percent of stream miles and 16 percent of lakes and reservoirs), with leading stream and river impairments being elevated levels of pathogens and nutrients.[32]

Waste

Currently, the world generates around 1.3 billion tonnes of municipal solid waste, including residential, industrial, commercial, institutional, municipal, and construction and demolition waste, annually, and this figure is expected to increase to approximately 2.2 billion tonnes per year by 2025. This represents a significant increase in per capita waste generation from 1.2 kg to 1.42 kg per person per day. Waste generation rates are influenced by economic development, the degree of industrialisation, public habits, and local climates.[33] Global solid waste generation is accelerating, particularly in urbanised areas. In 1900, 220 million people lived in cities, generating around 300,000 tonnes of rubbish per day. By 2000, the 2.9 billion people living in cities were generating more than 3 million tonnes (MT) of solid waste per day and by 2025 it will reach around 6 MT.[34] In India, economic growth and modern urban family living is increasing the amount of waste generated per capita, from 375 g/day in 1971 to a projected 700 g/day in 2025, while total urban municipal waste generation will increase from 14.9 MT/year in 1971 to nearly 100 MT/year in 2025.[35]

Air Pollution

Air pollution has emerged as one of the world's leading health risks. Each year more than 5.5 million people around the world die prematurely from illnesses caused by breathing polluted air. Air pollution is particularly severe in the world's fastest-growing urban areas where increased economic activity is contributing to higher levels of pollution and greater exposure.[36] Exposure to air pollution can vary significantly by socioeconomic status[37]; for example, in a study on the entire Medicare population in the United States, researchers found significant evidence of adverse effects related to exposure to $PM_{2.5}$ and ozone at concentrations below current national standards, with the effect most pronounced in self-identified racial minorities and people with low-income levels.[38] Overall, the OECD predicts that air pollution-related healthcare costs will increase from $21 billion in 2015 to $176 billion by 2060 due to a larger number of additional cases of illness and a projected increase in the healthcare costs per illness. By 2060, the annual number of air pollution-related lost working days, which impacts labour productivity, is expected to reach 3.7 billion, up from the current 1.2 billion.[39]

Erosion of Ecosystem Services

While humans receive numerous benefits from the natural environment in the form of goods and services (ecosystem services), including food, wood, clean water, energy, medicine, protection from floods and soil erosion as well as carbon storage, human development has also shaped the environment leading to ecosystem degradation across the globe. For example, over the past 300 years the global forest has shrunk by around 40 percent, since 1900 the world has lost around 50 percent of its wetlands, and the human-caused rate of species extinction is estimated to be 1000 times more rapid than the 'natural' rate.[40,41] The impact of these various trends is that approximately 60 percent of the Earth's ecosystem services have been degraded over the past 50 years. For instance, in New Zealand the value of ecosystem services provided by Lake Rotorua in 2012 was calculated to be NZD 94–138 million per annum, however, the costs of eutrophication are calculated to be $14–48 million per year. Overall, the values of ecosystems and potential losses associated with their degradation are often ignored in economic assessments.[42]

Climate Change

With rising greenhouse gas emissions, climate change is expected to lead to more frequent and severe flooding events, extended periods of heatwaves, and the loss of ecosystem services. Some of the impacts of climate change on the economy include changes in crop yields, loss of land and capital from sea level rise, increased mortality from heat stress, changes in energy demand for heating and cooling, and changes in availability of drinking water for end users including households and industry. In the United States, the combined value of market and non-market damages from climate change across a variety of sectors is around 1.2 percent of GDP per 1°C increase on average. By the late twenty-first century, the poorest third of countries are projected to experience damage between 2 and 20 percent of GDP under a business-as-usual emissions scenario.[43] The OECD predicts that in the absence of further action to address climate change, the combined negative effect on global annual GDP could be between 1 and 3.3 percent by 2060. For example, by 2050 climate change in the United States and European Union could lower annual economic growth by up to 1.90 percent and 2.17 percent respectively.[44]

Sustainable Development and Natural Capital

The linear economy is leading to rapid accumulation of physical and human capital while natural capital is being depleted and degraded, undermining the concept of sustainable development, which is defined by the Brundtland Commission as *"development that meets the needs of the present without compromising the ability of future generations to meet their own needs"*. This means future generations should be entitled to at least the same level of economic opportunities and economic welfare as currently available to present generations. Nonetheless, it is the total stock of capital employed by the economy, including natural capital, that determines the full range of economic opportunities and well-being for both present and future generations. As such, societies need to determine how best to use their available capital stock to increase current economic activities and welfare and how best to save or accumulate it for tomorrow for the well-being of future generations.[45] Therefore, it is not only the aggregate stock of capital in an economy that matters but also its composition.

In the weak form of sustainability, there is no difference between natural and other forms of capital: if depleted natural capital is being replaced with even more valuable physical and human capital, then the aggregate stock (human, physical, and the remaining natural) is increasing over time. In contrast, the strong sustainability view is that physical or human capital cannot substitute for all the environmental resources that comprise natural capital because there is uncertainty over many environmental values, such as the value that future generations may place on increasingly scarce natural resources and ecosystem services. This limits our ability to determine whether we can adequately compensate future generations for irreversible losses in natural capital today. As such, proponents of strong sustainability argue that environmental resources and ecosystem services essential for human welfare cannot be easily substituted by human and physical capital and so should be protected and not depleted.[46] Taking the strong sustainability view, there are many characteristics that distinguish natural capital from other forms of capital, therefore making it difficult for natural capital to be substituted, including

- *Exhaustibility*: Natural capital, unlike human and physical capital, cannot be produced by human activity, meaning if depleted or degraded it is often difficult to replace or restore. Non-renewable resource stocks cannot be regenerated once exploited or can only be replenished through natural cycles that are long in human-scale. Meanwhile, renewable resource stocks can regenerate over a relatively short period of time through natural processes, for example, freshwater and forests, but if over-exploited renewable resource stocks can be exhausted.
- *Geographic distribution*: Natural resources are often unevenly distributed across the world. Trade has helped alleviate some of these disparities by allowing natural resources to be redistributed from countries with excess supply to countries with limited or no domestic supplies.
- *Availability and accessibility*: The availability of materials required to support economic activity depends on both a country's endowment in natural resources and on the accessibility of external sources of material inputs (via imports). Whether availability is a constraint on economic growth or not depends on the efficiency of resource use and the capacity of the economy to innovate and develop new technologies. At the international level, accessibility is influenced by geographical distribution, worldwide demands, and market prices for materials, as well as political, regulatory, and institutional factors.
- *Externalities and pricing*: Natural resources are associated with both positive and negative externalities, resulting in economic undervaluation of natural capital. When natural resources are used as economic inputs, they generate a commercial return; however, their non-use benefits cannot be captured and therefore are not reflected in market prices, and the production and consumption of natural resources can have a negative effect on the environment including air and water pollution and waste.[47]

Overall, the strong sustainability view calls for an economy that is based on a 'borrow-use-replenish framework' where natural resources are converted into goods and services with the by-products returned to the economy or returned to nature as nutrients for further use, without degrading the structure and function of the natural environment.[48]

The Circular Economy

The circular economy, in contrast to the linear 'take-make-consume-dispose' economy, aims to keep resources in use for as long as possible, extract value from them while in use, and recover and regenerate products and materials at the end of each service life.[49] As such, the circular economy focuses on recycling, limiting, and reusing the physical inputs of the economy and using waste as a resource, leading to reduced primary resource consumption.[50] This is commonly referred to as the 3R (reduce, reuse, and recycle) approach. A key aspect of this approach is that materials, which have accumulated in the circular economy, constitute important man-made stocks that can be exploited through recycling to gain secondary raw materials and reused and remanufactured to keep products in the commercial life cycle.[51]

Decoupling Economic Growth from Resource Use

The aim of the circular economy is to decouple economic growth from resource use and associated environmental impacts. The notion of decoupling is that economic output shall continue to increase at the same time as rates of increasing resource use and environmental impact are slowed, and in time brought into decline.[52] Specifically, decoupling is said to be absolute when the environmental variable is stable or decreasing while the economic variable is growing. Decoupling is considered relative when the environmental variable is increasing but at a lower rate than the economic variable. In the context of natural resources management, two modes of decoupling can be distinguished:

- Resource decoupling or 'dematerialisation' involves reducing the rate at which natural resources are used per unit of economic output
- Impact decoupling seeks to increase economic activity while decreasing negative environmental impacts from pressures such as pollution, carbon emissions, or destruction of biodiversity[53]

Product Life Cycles in the Circular Economy

A product's life cycle can be broken down into stages that flow from upstream to downstream,[54] including:

- Product design
- Raw material extraction and processing
- Manufacturing of the product
- Packaging and distribution to the consumer
- Product use and maintenance
- End-of-life management (reuse, recycling, and disposal)

In the circular economy, waste is 'designed out' of a product's life cycle through innovation (summarised in Table 1.1), rather than relying on solutions at the end of a product's life.[55] As such, an economic actor that wishes to design out waste in a new product will assess how both the design phase and production processes have an impact on sourcing, resource use, and waste generation throughout a product's life cycle (Fig. 1.1).[56]

Key Differences Between the Linear and Circular Economies

The key differences between the linear and circular economies are summarised in Table 1.2 which provides a general overview of the key mechanisms that shape the role and fate of products in both linear and circular economies seen through the lens of businesses selling products, consumers buying and using products, and policymakers regulating the production, use, and end of life of products.[57]

Table 1.1 Innovative circular economy approaches

Circular economy approaches	Description
Light-weighting	Reducing the quantity of materials required to deliver a service
Durability	Lengthening a product's useful life
Efficiency	Reducing the use of energy and materials in production and use phases
Substitution	Reducing the use of materials that are hazardous or difficult to recycle in products and production processes
Recyclates	Creating markets for secondary raw materials
Eco-design	Designing products that are easier to maintain, repair, upgrade, remanufacture, or recycle
Maintenance/ repair services	Developing the necessary services for consumers to have products maintained or repaired
Waste reduction	Incentivising and supporting waste reduction and high-quality separation by consumers
Waste separation	Incentivising separation and collection systems that minimise the costs of recycling and reuse
Industrial symbiosis	Facilitating the clustering of activities to prevent by-products from becoming wastes
Consumer options	Encouraging wider and better consumer choice through renting, lending, or sharing services as alternatives to owning products, while safeguarding consumer interests (in terms of costs, protection, information, contract terms, insurance aspects, etc.)

The Role of Government in Developing a Circular Economy

Government intervention has an important role in developing the circular economy and encouraging a life cycle perspective to be taken by economic actors.[58] Some of the key aspects governments need to consider in encouraging the development of a circular economy are as follows.

Encouraging Better Product Design

Better design can make products more durable or easier to repair, upgrade, or remanufacture. It can help recyclers disassemble products to recover valuable materials or components, which overall reduces resource use. However, current market signals are insufficient for this to happen and

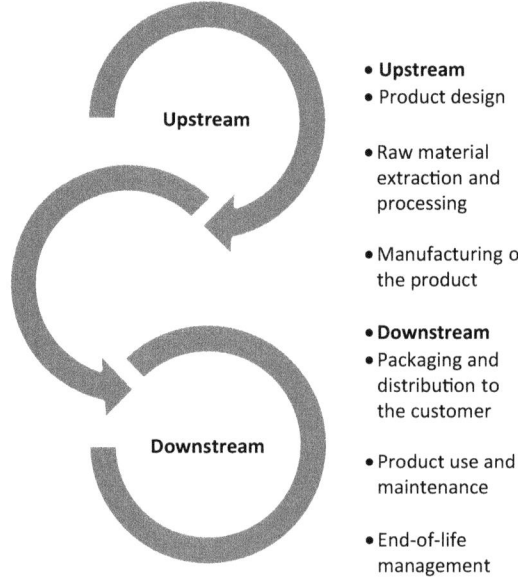

Upstream

Downstream

- **Upstream**
- Product design

- Raw material extraction and processing

- Manufacturing of the product

- **Downstream**
- Packaging and distribution to the customer

- Product use and maintenance

- End-of-life management

Fig. 1.1 A product's life cycle

therefore it is essential to provide incentives for improved product design while preserving competition and innovation. Nonetheless, even for products designed in a 'smart' way, inefficient use of resources in production processes can lead to significant costs and waste generation. As such, industry has a role in sustainably sourcing raw materials and cooperating across value chains. While each industry sector is different in terms of resource use, waste generation, and management, governments can promote best practices, help small and medium-sized enterprises to benefit from the business opportunities that arise from resource efficiency, as well as promote innovative industrial processes.[59]

Facilitating Better Consumption Choices

The choices consumers make can support or hamper the development of the circular economy. These choices are shaped by the information con-

Table 1.2 Differences between the linear and circular economy models

Linear economy	Circular economy
Business perspective	
Products as value creation sources Profit margins are based on the difference between the market price of a product and the production cost. The strategy of increasing profits is to sell more products and keep production costs as low as possible. Technological innovation makes old products obsolete and urges customers to buy new products	**Functionality/performance as a source of value creation** Products are part of an integrated business model that focuses on the delivery of a performance or functional service. Competition is mainly based on the creation of added service value of a product. Social/business model innovation allows the creation of extra value by applying technological innovation to solving societal needs
Economics of scale in global production chains Cost efficiency drives the optimisation of global production chains, minimising the costs of resources, labour, and transport	**Location of production and use tend to be more linked** As the provision of a service is physically linked to the location of a customer, there is an incentive to produce/manage physical products used in a service close to the user
Steer consumer needs towards product offer Products with short lifespans are preferred as they are cheaper to make and support a market for new products that replace old ones	**User needs/wants drive the role of a product** Offering the best service means matching the needs of the user with a combination of services and products
Tendency to disregard end-of-life phase There is no economic incentive for product life extension, reuse, or remanufacturing as they counteract most linear business models	**Internal incentive to incorporate end-of-life phase in business model** Minimising life cycle costs is an incentive for a company
Consumer perspective	
Consumerism follows marketing Consumers want new products to keep pace with fashion and technological advances	**Customer satisfaction is an important driver** The customer experience feeds back to the service provider, raising customer awareness of their actual needs

(continued)

Table 1.2 (continued)

Linear economy	Circular economy
International opportunities for cost reduction	**Local-first attitude**
Consumers seek the cheapest version of a product on international markets, enabled by e-commerce	Accessibility to the service provider is part of the service experience, and so proximity is a customer choice criterion
Ownership is the norm	**Accessibility is the norm**
Owning a product is 'normal' in fulfilling needs. Over time, luxury goods become commodity goods. Product repair is too expensive compared to buying a new product	Fulfilling needs is driven by accessibility of a product and the satisfaction provided by its use. Different consumer segments can access products of their choice through customised services or by sharing products. Service agreements provide incentives for product care
Low/no residual value of products	**End-of-use incorporated**
End-of-life products (broken or obsolete) are a burden to be disposed of as cheaply as possible, by selling, storing at home, or through regulated waste disposal systems or illegal incineration/dumping	If products are part of a service, there are incentives to return them to the provider after use, avoiding stocks of obsolete products in households or illegal dumping
Policy perspective	
Dependence of existing production system	**More focus on facilitating skilled workforce**
There is a strong link between mass production of goods and the focus on cutting costs and making production as efficient as possible, often resulting in lower labour costs and less job creation	More localised and service-based activities require a skilled but affordable workforce
Global playing field	**Less risk for outsourcing jobs**
Competition for economic factors on the international market steers national social and environmental policies	As production uses local assets, there is less likelihood of outsourcing, removing the incentive for 'race-to-the-bottom' social and environmental policies
Balance consumer protection with economic stakes	**Facilitate safe and healthy services with regulation**
Protection of consumer health and safety is mainly reactive and geared towards protecting existing economic stakes, e.g. value-added tax income	Because consumer health and safety are business incentives for high-quality performance, policies focus on these types of services

(continued)

Table 1.2 (continued)

Linear economy	Circular economy
Action prompted by health or environmental concerns	**Facilitation of end-of-life management**
There is no inherent incentive for regulation of the waste phase of products. Only when waste-related health or environmental concerns arise is regulatory action taken to minimise negative impacts	Extended producer responsibility rules create incentives for companies to internalise end-of-life management. Governments provide basic infrastructure and fiscal measures to support reverse logistics

sumers have access to, the range and prices of existing products, and the regulatory environment. Price is a key factor affecting purchasing decisions, both in the value chain and for final consumption. Therefore, governments need to provide incentives and use economic instruments to ensure that product prices better reflect environmental costs. Actions can be taken to reduce the amount of household waste, for example, initiating awareness campaigns and implementing economic incentives.[60]

Improving Waste Management

Waste management plays a central role in the circular economy. The way that waste is collected and managed determines whether there are high levels of recycling or not and whether valuable materials find their way back into the economy, or to an inefficient system where most recyclable waste ends up in landfills or is incinerated, with potentially harmful environmental costs and significant economic losses. To increase levels of high-quality recycling, improvements need to be made in waste collection and sorting, such as ensuring adequate investment in separate collection and recycling infrastructure and use of economic instruments to encourage recycling.

Creating a Market for Waste to Resources

In the circular economy, materials can be recycled back into the economy as new raw materials, thereby increasing the security of supply. Secondary raw materials can be used just like primary raw materials from traditional extractive resources. A key factor in creating a dynamic market for secondary raw materials is sufficient demand. Governments can contribute to this demand for recycled materials through their procurement policies.[61]

Guidance on Developing the Circular Economy

There are four broad principles that should guide the development of circular economy policies that encourage a life cycle perspective to be taken by economic actors wherever possible.

1. *Preserve natural capital*: Natural resources and healthy ecosystems are essential to all life—human and natural—and provide the natural capital that humans depend on. The circular economy can contribute to the preservation of natural capital by leveraging science, engineering, business, and management practices to halt and reverse the depletion of natural capital. Policies can be developed to:

   - Improve information about material flows and environmental impacts
   - Increase resource productivity and resource efficiency
   - Reduce material throughput
   - Increase reuse/recycling of materials to preserve natural capital
   - Advance technologies for obtaining materials from natural resources that eliminate waste and toxins and support long-term ecosystem health

2. *Take a life cycle perspective*: It is at the design stage that decisions are made that determine impacts throughout the life cycle of a product. Policies can be developed to maximise positive impacts to the environment and to human health and well-being through design. Managing safety and sustainability at each life cycle stage ensures that risks are not shifted from one stage in the value chain, or from one geographical area, to another. Three main material, product, and process design strategies support the circular economy and can be encouraged via government policies:

1. Detoxification supports the elimination of chemicals and compounds produced by society that have harmful impacts on human health and the environment, cannot be properly or safely managed, or are costly to manage from an economic or environmental perspective.
2. Dematerialisation supports the circular economy by reducing the throughput of materials, particularly those with high negative life cycle impacts. Dematerialisation means doing more with less and refers to the efficient use of raw materials (resource efficiency) without decreasing the quality of the service they provide.
3. Design for value recovery supports the circular economy by ensuring that products and materials are designed for reuse and recycling and that an effective model for recovery is in place.

3. *Use policy instruments to achieve sustainable economic and environmental outcomes*: To shift the economy from a linear to a circular model, governments can leverage a variety of policies and policy tools including regulations, economic incentives, innovation policies, information sharing, and partnerships. It is unlikely that one single mechanism is appropriate in all circumstances and therefore a multi-pronged approach, applying a diverse range of policies and policy tools, is more likely to influence all relevant stakeholders than a 'one-size-fits-all' approach. Policy tool combinations can reinforce each other to help generate more effective, efficient, and lasting outcomes.
4. *Engage stakeholders to achieve sustainable outcomes*: Material flows involve and affect many stakeholders throughout the supply chain and across geographical areas. As such, outcomes can be improved by inclusion and engagement of many players across the life cycle of materials through collaborative efforts that lead to collective solutions. Stakeholder engagement can also facilitate socially acceptable and equitable solutions by engaging those affected and allowing them to participate in designing solutions. Outcomes can be improved through:

- Multilateral stakeholder engagement, responsibility, and collaboration
- Open information flows
- An ethical perspective[62]

Notes

1. Ellen MacArthur Foundation. 2013. Towards the circular economy: Economic and business rationale for an accelerated transition. 1. Available: https://www.ellenmacarthurfoundation.org/assets/downloads/publications/Ellen-MacArthur-Foundation-Towards-the-Circular-Economy-vol.1.pdf.
2. Barbier, E. B. 2003. The role of natural resources in economic development. *Australian Economic Papers,* 42, 253–272.
3. Goodwin, N. H., Jonathan; Nelson, Julie A.; Roach, Brian; and Torras, Mariano 2013. *Microeconomics in context,* Armonk, New York, M.E. Sharpe.
4. World Economic Forum. 2014. Towards the Circular Economy: Accelerating the scale-up across global supply chains. Available: http://www3.weforum.org/docs/WEF_ENV_TowardsCircularEconomy_Report_2014.pdf.
5. OECD. 2015b. Material resources, productivity and the environment. Available: http://www.oecd.org/env/waste/material-resources-productivity-and-the-environment-9789264190504-en.htm.
6. PWC. 2015. The world in 2050. Will the shift in global economic power continue? Available: https://www.pwc.com/gx/en/issues/the-economy/assets/world-in-2050-february-2015.pdf.
7. UNEP. 2015. International trade in resources: A biophysical assessment. Available: http://www.resourcepanel.org/international-trade-resources-biophysical-assessment.
8. European Commission. 2011. Roadmap to a resource efficient Europe. Available: http://eur-lex.europa.eu/legal-content/EN/TXT/PDF/?uri=CELEX:52011DC0571&from=EN.
9. WHO. 2017. *Global and regional food consumption patterns and trends* [Online]. Available: http://www.who.int/nutrition/topics/3_foodconsumption/en/index4.html.
10. HSBC. 2012. Consumer in 2050. The rise of the EM middle class. Available: https://www.hsbc.com.vn/1/PA_ES_Content_Mgmt/content/vietnam/abouthsbc/newsroom/attached_files/HSBC_report_Consumer_in_2050_EN.pdf.
11. Future Market Insights. 2016. *Consumer electronics to be a US$ 3 trillion market by 2020: Report* [Online]. Available: http://www.futuremarketinsights.com/press-release/consumer-electronics-market.

12. UNEP. 2016. Resource efficiency: Potential and economic implications. Available: http://www.resourcepanel.org/sites/default/files/documents/document/media/resource_efficiency_report_march_2017_web_res.pdf.
13. Henckens, M., Van Ierland, E., Driessen, P. & Worrell, E. 2016. Mineral resources: Geological scarcity, market price trends, and future generations. *Resources Policy,* 49, 102–111.
14. European Commission. 2010. *Report forecasts shortages of 14 critical mineral raw materials* [Online]. Available: http://europa.eu/rapid/press-release_IP-10-752_en.htm.
15. Mckinsey and Company. 2011. Resource revolution: Meeting the world's energy, materials, food, and water needs. Available: http://www.mckinsey.com/business-functions/sustainability-and-resource-productivity/our-insights/resource-revolution.
16. UNEP. 2016. Resource efficiency: Potential and economic implications. Available: http://www.resourcepanel.org/sites/default/files/documents/document/media/resource_efficiency_report_march_2017_web_res.pdf.
17. Bekkers, E., Brockmeier, M., Francois, J. & Yang, F. 2017. Local Food Prices and International Price Transmission. *World Development,* 96, 216–230.
18. UN. 2017. World population prospects. The 2017 revision. Available: https://esa.un.org/unpd/wpp/Publications/Files/WPP2017_KeyFindings.pdf.
19. UN. 2014. World urbanization prospects: The 2014 revision, highlights. Available: https://esa.un.org/unpd/wup/publications/files/wup2014-highlights.Pdf.
20. Bounoua, L., Zhang, P., Nigro, J., Lachir, A. & Thome, K. 2017. Regional Impacts of Urbanization in the United States. *Canadian Journal of Remote Sensing,* 43, 256–268.
21. PRB. 2004. *Urbanization: An environmental force to be reckoned with* [Online]. Available: http://www.prb.org/Publications/Articles/2004/UrbanizationAnEnvironmentalForcetoBeReckonedWith.aspx
22. Cui, L. & Shi, J. 2012. Urbanization and its environmental effects in Shanghai, China. *Urban Climate,* 2, 1–15.
23. Mckinsey and Company. 2011. Resource revolution: Meeting the world's energy, materials, food, and water needs. Available: http://www.mckinsey.com/business-functions/sustainability-and-resource-productivity/our-insights/resource-revolution.

24. PWC. 2017. *Capital project and infrastructure spending outlook* [Online]. Available: https://www.pwc.com/gx/en/industries/capital-projects-infrastructure/publications/cpi-spending-outlook.html.
25. Hughes, G., Chinowsky, P. & Strzepek, K. 2010. The costs of adaptation to climate change for water infrastructure in OECD countries. *Utilities Policy,* 18, 142–153.
26. EIA. 2016b. International energy outlook 2016. Available: https://www.eia.gov/outlooks/ieo/world.php.
27. EIA. 2016a. *EIA projects 48% increase in world energy consumption by 2040* [Online]. Available: https://www.eia.gov/todayinenergy/detail.php?id=26212.
28. Bao, C., Kishita, Y. & Umeda, Y. 2017. Demand estimation of consumer durables in Southeast Asia in 2030: A business-as-usual scenario. *Procedia CIRP,* 61, 635–640.
29. UNESCO. 2015. Water for a sustainable world. Facts and figures. Available: http://www.unesco.org/fileadmin/MULTIMEDIA/HQ/SC/images/WWDR2015Facts_Figures_ENG_web.pdf.
30. OECD. 2012a. Environmental outlook to 2050: The consequences of inaction. Key findings of water. Available: https://www.oecd.org/env/indicators-modelling-outlooks/49844953.pdf.
31. UN-WATER. 2017. Wastewater: The untapped resource. Available: http://www.unesco.org/new/en/natural-sciences/environment/water/wwap/.
32. Braden, J. B. & Shortle, J. S. 2013. Agricultural sources of water pollution A2 – Shogren, Jason F. *Encyclopedia of energy, natural resource, and environmental economics.* Waltham: Elsevier.
33. World Bank. 2017. What a waste: A global review of solid waste management. Available: https://openknowledge.worldbank.org/handle/10986/17388.
34. Hoornweg, D., Bhada-Tata, P. & Kennedy, C. 2013. Environment: Waste production must peak this century. *Nature,* 502, 615.
35. Vij, D. 2012. Urbanization and solid waste management in India: Present practices and future challenges. *Procedia – Social and Behavioral Sciences,* 37, 437–447.
36. World Bank. 2016. The cost of air pollution: Strengthening the economic case for action. Available: https://openknowledge.worldbank.org/handle/10986/25013.
37. NIH. 2016. *Poor communities exposed to elevated air pollution levels* [Online]. Available: https://www.niehs.nih.gov/research/programs/geh/

geh_newsletter/2016/4/spotlight/poor_communities_exposed_to_elevated_air_pollution_levels.cfm.

38. Di, Q., Wang, Y., Zanobetti, A., Wang, Y., Koutrakis, P., Choirat, C., Dominici, F. & Schwartz, J. D. 2017. Air Pollution and Mortality in the Medicare Population. *New England Journal of Medicine,* 376, 2513–2522.

39. OECD. 2015a. The economic consequences of outdoor air pollution. Available: http://www.oecd.org/environment/indicators-modelling-outlooks/the-economic-consequences-of-outdoor-air-pollution-9789264257474-en.htm.

40. Millennium Ecosystem Assessment. 2005. *Ecosystems and human wellbeing: Synthesis.* Washington, DC, Island Press.

41. TEEB. 2010. *The economics of ecosystems and biodiversity: Mainstreaming the economics of nature: A synthesis of the approach, conclusions and recommendations of TEEB* [Online]. Available: http://www.teebweb.org/our-publications/teeb-study-reports/synthesis-report/.

42. Mueller, H., Hamilton, D. P. & Doole, G. J. 2016. Evaluating services and damage costs of degradation of a major lake ecosystem. *Ecosystem Services,* 22, Part B, 370–380.

43. Hsiang, S., Kopp, R., Jina, A., Rising, J., Delgado, M., Mohan, S., Rasmussen, D. J., Muir-Wood, R., Wilson, P., Oppenheimer, M., Larsen, K. & Houser, T. 2017. Estimating economic damage from climate change in the United States. *Science,* 356, 1362–1369.

44. Du, D., Zhao, X. & Huang, R. 2017. The impact of climate change on developed economies. *Economics Letters,* 153, 43–46.

45. Barbier, E. B. 2003. The Role of Natural Resources in Economic Development. *Australian Economic Papers,* 42, 253–272.

46. Ibid.

47. OECD. 2015c. Material resources, productivity and the environment. *OECD Green Growth Studies* [Online]. Available: http://www.oecd.org/env/waste/material-resources-productivity-and-the-environment-9789264190504-en.htm.

48. Pike, C., Doppelt, B. & Herr, M. 2010. Climate communications and behavior change: A guide for practitioners. Available: http://static1.1.sqspcdn.com/static/f/551504/6527501/1271194957847/Climate+Communications+and+Behavior+Change.pdf?token=Eo1SlEShO gUlVMqJBMuyPLvVnuo%3D.

49. WRAP. 2017. *WRAP and the circular economy* [Online]. Available: http://www.wrap.org.uk/about-us/about/wrap-and-circular-economy.

50. EEA. 2014. Resource-efficient green economy and EU policies. Available: http://www.eea.europa.eu/publications/resourceefficient-green-economy-and-eu.

51. OECD. 2015c. Material resources, productivity and the environment. *OECD Green Growth Studies* [Online]. Available: http://www.oecd.org/env/waste/material-resources-productivity-and-the-environment-9789264190504-en.htm.

52. UNEP. 2016. Resource efficiency: Potential and economic implications. Available: http://www.resourcepanel.org/sites/default/files/documents/document/media/resource_efficiency_report_march_2017_web_res.pdf.

53. OECD. 2015c. Material resources, productivity and the environment. *OECD Green Growth Studies* [Online]. Available: http://www.oecd.org/env/waste/material-resources-productivity-and-the-environment-9789264190504-en.htm.

54. Gupt, Y. & Sahay, S. 2015. Review of extended producer responsibility: A case study approach. *Waste Management & Research,* 33, 595–611.

55. European Commission. 2014. Towards a circular economy: A zero waste programme for Europe. Available: http://eur-lex.europa.eu/resource.html?uri=cellar:aa88c66d-4553-11e4-a0cb-01aa75ed71a1.0022.03/DOC_1&format=PDF.

56. UNEP. 2005. Life cycle approaches. The road from analysis to practice. Available: http://www.unep.fr/shared/publications/pdf/DTIx0594xPA-Road.pdf.

57. EEA. 2017. Circular by design – Products in the circular economy. Available: https://www.eea.europa.eu/publications/circular-by-design.

58. Way, T. K., Ong, M., Kai, J., Ho, S. & Michelle Kan 2016. Is your waste a waste? Rethinking the linear economy. *Asian Management Insights,* 3, 62–69.

59. European Commission. 2015. Closing the loop – An EU action plan for the Circular Economy. Available: http://eur-lex.europa.eu/legal-content/EN/TXT/?uri=CELEX%3A52015DC0614.

60. Ibid.

61. Ibid.

62. OECD. 2012b. Sustainable materials management. Making better use of resources. Available: https://doi.org/10.1787/9789264174269-en.

References

BAO, C., Y. Kishita, and Y. Umeda. 2017. Demand estimation of consumer durables in Southeast Asia in 2030: A business-as-usual scenario. *Procedia CIRP* 61: 635–640.

Barbier, E.B. 2003. The role of natural resources in economic development. *Australian Economic Papers* 42: 253–272.

Bekkers, E., M. Brockmeier, J. Francois, and F. Yang. 2017. Local food prices and international price transmission. *World Development* 96: 216–230.

Bounoua, L., P. Zhang, J. Nigro, A. Lachir, and K. Thome. 2017. Regional impacts of urbanization in the United States. *Canadian Journal of Remote Sensing* 43: 256–268.

Braden, J.B., and J.S. Shortle. 2013. Agricultural sources of water pollution A2. In *Encyclopedia of energy, natural resource, and environmental economics*, ed. Jason F. Shogren. Waltham: Elsevier.

Cui, L., and J. Shi. 2012. Urbanization and its environmental effects in Shanghai, China. *Urban Climate* 2: 1–15.

Di, Q., Y. Wang, A. Zanobetti, Y. Wang, P. Koutrakis, C. Choirat, F. Dominici, and J.D. Schwartz. 2017. Air pollution and mortality in the medicare population. *New England Journal of Medicine* 376: 2513–2522.

Du, D., X. Zhao, and R. Huang. 2017. The impact of climate change on developed economies. *Economics Letters* 153: 43–46.

EEA. 2014. Resource-efficient green economy and EU policies. Available: http://www.eea.europa.eu/publications/resourceefficient-green-economy-and-eu.

———. 2017. Circular by design – Products in the circular economy. Available: https://www.eea.europa.eu/publications/circular-by-design.

EIA. 2016a. *EIA projects 48% increase in world energy consumption by 2040* [Online]. Available: https://www.eia.gov/todayinenergy/detail.php?id=26212.

———. 2016b. International energy outlook 2016. Available: https://www.eia.gov/outlooks/ieo/world.php.

Ellen MacArthur Foundation. 2013. Towards the circular economy: Economic and business rationale for an accelerated transition. 1. Available: https://www.ellenmacarthurfoundation.org/assets/downloads/publications/Ellen-MacArthur-Foundation-Towards-the-Circular-Economy-vol.1.pdf.

European Commission. 2010. *Report forecasts shortages of 14 critical mineral raw materials* [Online]. Available: http://europa.eu/rapid/press-release_IP-10-752_en.htm.

————. 2011. Roadmap to a resource efficient Europe. Available: http://eur-lex.europa.eu/legal-content/EN/TXT/PDF/?uri=CELEX:52011DC0571&from=EN.

————. 2014. Towards a circular economy: A zero waste programme for Europe. Available: http://eur-lex.europa.eu/resource.html?uri=cellar:aa88c66d-4553-11e4-a0cb-01aa75ed71a1.0022.03/DOC_1&format=PDF.

————. 2015. Closing the loop – An EU action plan for the Circular Economy. Available: http://eur-lex.europa.eu/legal-content/EN/TXT/?uri=CELEX%3A52015DC0614.

Future Market Insights. 2016. *Consumer electronics to be a US$ 3 trillion market by 2020: Report* [Online]. Available: http://www.futuremarketinsights.com/press-release/consumer-electronics-market.

Goodwin, N., H. Jonathan, Julie A. Nelson, Brian Roach, and Mariano Torras. 2013. *Microeconomics in context*. Armonk: M.E. Sharpe.

Gupt, Y., and S. Sahay. 2015. Review of extended producer responsibility: A case study approach. *Waste Management & Research* 33: 595–611.

Henckens, M., E. Van Ierland, P. Driessen, and E. Worrell. 2016. Mineral resources: Geological scarcity, market price trends, and future generations. *Resources Policy* 49: 102–111.

Hoornweg, D., P. Bhada-Tata, and C. Kennedy. 2013. Environment: Waste production must peak this century. *Nature* 502: 615.

HSBC. 2012. Consumer in 2050. The rise of the EM middle class. Available: https://www.hsbc.com.vn/1/PA_ES_Content_Mgmt/content/vietnam/abouthsbc/newsroom/attached_files/HSBC_report_Consumer_in_2050_EN.pdf.

Hsiang, S., R. Kopp, A. Jina, J. Rising, M. Delgado, S. Mohan, D.J. Rasmussen, R. Muir-Wood, P. Wilson, M. Oppenheimer, K. Larsen, and T. Houser. 2017. Estimating economic damage from climate change in the United States. *Science* 356: 1362–1369.

Hughes, G., P. Chinowsky, and K. Strzepek. 2010. The costs of adaptation to climate change for water infrastructure in OECD countries. *Utilities Policy* 18: 142–153.

McKinsey and Company. 2011. Resource revolution: Meeting the world's energy, materials, food, and water needs. Available: http://www.mckinsey.com/business-functions/sustainability-and-resource-productivity/our-insights/resource-revolution.

Millennium Ecosystem Assessment. 2005. *Ecosystems and human well-being: Synthesis*. Washington, DC: Island Press.

Mueller, H., D. P. Hamilton, and G. J. Doole. 2016. Evaluating services and damage costs of degradation of a major lake ecosystem. *Ecosystem Services* 22, Part B: 370–380.

NIH. 2016. *Poor communities exposed to elevated air pollution levels* [Online]. Available: https://www.niehs.nih.gov/research/programs/geh/geh_newsletter/2016/4/spotlight/poor_communities_exposed_to_elevated_air_pollution_levels.cfm.

OECD. 2012a. Environmental outlook to 2050: The consequences of inaction. Key findings of water. Available: https://www.oecd.org/env/indicators-modelling-outlooks/49844953.pdf.

———. 2012b. Sustainable materials management. Making better use of resources. Available: https://doi.org/10.1787/9789264174269-en.

———. 2015a. The economic consequences of outdoor air pollution. Available: http://www.oecd.org/environment/indicators-modelling-outlooks/the-economic-consequences-of-outdoor-air-pollution-9789264257474-en.htm.

———. 2015b. The economics of climate change. Available: https://doi.org/10.1787/9789264235410-en.

———. 2015c. Material resources, productivity and the environment. Available: http://www.oecd.org/env/waste/material-resources-productivity-and-the-environment-9789264190504-en.htm.

———. 2015d. Material resources, productivity and the environment. *OECD Green Growth Studies* [Online]. Available: http://www.oecd.org/env/waste/material-resources-productivity-and-the-environment-9789264190504-en.htm.

Pike, C., B. Doppelt, and M. Herr. 2010. Climate communications and behavior change: A guide for practitioners. Available: http://static1.1.sqspcdn.com/static/f/551504/6527501/1271194957847/Climate+Communications+and+Behavior+Change.pdf?token=Eo1SlESh0gUlVMqJBMuyPLvVnuo%3D.

PRB. 2004. *Urbanization: An environmental force to be reckoned with* [Online].

PWC. 2015. The world in 2050. Will the shift in global economic power continue? Available: https://www.pwc.com/gx/en/issues/the-economy/assets/world-in-2050-february-2015.pdf.

———. 2017. *Capital project and infrastructure spending outlook* [Online]. Available: https://www.pwc.com/gx/en/industries/capital-projects-infrastructure/publications/cpi-spending-outlook.html.

TEEB. 2010. *The economics of ecosystems and biodiversity: Mainstreaming the economics of nature: A synthesis of the approach, conclusions and recommendations of TEEB* [Online]. Available: http://www.teebweb.org/our-publications/teeb-study-reports/synthesis-report/.

UN-Water. 2017. Wastewater: The untapped resource. Available: http://www.unesco.org/new/en/natural-sciences/environment/water/wwap/.

UN. 2014. World urbanization prospects: The 2014 revision, highlights. Available: https://esa.un.org/unpd/wup/publications/files/wup2014-highlights.Pdf.

———. 2017. World population prospects. The 2017 revision. Available: https://esa.un.org/unpd/wpp/Publications/Files/WPP2017_KeyFindings.pdf.

UNEP. 2005. Life cycle approaches. The road from analysis to practice. Available: http://www.unep.fr/shared/publications/pdf/DTIx0594xPA-Road.pdf.

———. 2015. International trade in resources: A biophysical assessment. Available: http://www.resourcepanel.org/international-trade-resources-biophysical-assessment.

———. 2016. Resource efficiency: Potential and economic implications. Available: http://www.resourcepanel.org/sites/default/files/documents/document/media/resource_efficiency_report_march_2017_web_res.pdf.

UNESCO. 2015. Water for a sustainable world. Facts and figures. Available: http://www.unesco.org/fileadmin/MULTIMEDIA/HQ/SC/images/WWDR2015Facts_Figures_ENG_web.pdf.

VIJ, D. 2012. Urbanization and solid waste management in India: Present practices and future challenges. *Procedia – Social and Behavioral Sciences* 37: 437–447.

Way, T.K., M. Ong, J. Kai, S. Ho, and Michelle Kan. 2016. Is your waste a waste? Rethinking the linear economy. *Asian Management Insights* 3: 62–69.

WHO. 2017. *Global and regional food consumption patterns and trends* [Online]. Available: http://www.who.int/nutrition/topics/3_foodconsumption/en/index4.html.

World Bank. 2016. The cost of air pollution: Strengthening the economic case for action. Available: https://openknowledge.worldbank.org/handle/10986/25013.

———. 2017. What a waste: A global review of solid waste management. Available: https://openknowledge.worldbank.org/handle/10986/17388.

World Economic Forum. 2014. Towards the Circular Economy: Accelerating the scale-up across global supply chains. Available: http://www3.weforum.org/docs/WEF_ENV_TowardsCircularEconomy_Report_2014.pdf.

WRAP. 2017. *WRAP and the circular economy* [Online]. Available: http://www.wrap.org.uk/about-us/about/wrap-and-circular-economy.

2

Circular Economy: Fiscal and Non-Fiscal Tools

Introduction

Most of the time environmental pollution and natural resource depletion come from incorrect pricing of the goods and services society produces and consumes. Fiscal tools help realise the circular economy by ensuring economic actors consider the 'hidden' costs of production and consumption, such as air and waste pollution, waste disposal, soil and species loss, and climate change, in a cost-effective way. Overall, fiscal tools provide a stimulus to producers and consumers to change their behaviour towards a more eco-efficient use of natural resources by stimulating technological innovation and reducing consumption levels.[1,2] Governments at all levels can also use a variety of non-fiscal tools to promote the development and uptake of circular economy-related technologies and services by modifying the attitudes and behaviour of producers and consumers towards natural resources.[3]

© The Author(s) 2018
R. C. Brears, *Natural Resource Management and the Circular Economy*,
Palgrave Studies in Natural Resource Management,
https://doi.org/10.1007/978-3-319-71888-0_2

Fiscal Tools

Governments at all levels can use a variety of fiscal tools to develop the circular economy and encourage a life cycle perspective to be taken by economic actors in an attempt to decouple economic growth from resource use and associated environmental impacts.

Environmental Taxes and Charges

Without government intervention, there is no market incentive for economic actors to consider environmental damage, as its impact is spread across many people and has little or no direct cost to the polluter. In the past, environmental policy was often dominated by 'command-and-control' regulations. This approach was generally prescriptive and highly targeted, for example, banning or limiting the use of specific substances or requiring certain industries to use specific technologies. Over recent decades there has been a shift towards environmental taxes and charges, where an environmental tax is a tax whose base is a physical unit (or a proxy of a physical unit) of something that has a proven, specific negative impact on the environment,[4] with typical environmental taxes being energy taxes, transport taxes, pollution taxes, and resources taxes.[5] Meanwhile, environmental charges are paid by economic actors for the right to use the environment. The purpose of imposing environmental charges is to prevent or reduce possible harm relating to the use of natural resources, prevent or reduce emissions of pollutants into the environment, and encourage the proper disposal of waste. The key benefits of governments using environmental taxes and charges rather than command-and-control regulations include:

- *Addressing market failure by pricing in environmental costs*: Environmental taxes and charges ensure environmental impacts are incorporated into prices. As such, a well-designed environmental tax/charge increases the price of a good or activity to reflect the cost of the environmental harm that it imposes on others. The cost of this harm to others—an externality—is internalised into market prices. This ensures economic actors take these costs into account in their decisions.

- *Providing economic actors with the flexibility to determine how best to reduce their environmental footprint:* Most regulatory approaches involve the government specifying how to reduce pollution. In contrast, environmental taxes and charges provide the flexibility in determining which goods or practices are most suitable in reducing environmental harm. This reduces the involvement of government in steering the economy in favour of certain environmental solutions over others.[6]

The flexibility of response associated with environmental taxes and charges also provides other benefits, such as:

- *Ongoing incentives to abate:* A target-based or technology-based regulation provides no incentive to abate once the target or technology standard is met. In contrast, environmental taxes and charges provide a continuous incentive to abate at all levels of emissions, even after significant abatement has occurred.
- *Improved competitiveness of low-polluting alternatives:* Environmental taxes and charges increase demand for low-polluting alternatives. This increases the economies of scale that helps make alternatives more viable without the need for direct subsidies.
- *Strong incentives to innovate:* Taxes increase the cost to a polluter of generating pollution, providing incentives for firms to develop new innovations and adapt existing ones.
- *Transparency:* Well-designed environmental taxes and charges are highly transparent in terms of their coverage and costs. It is clear what is taxed/charged, which polluters are exempt, and what the costs to polluters will be per unit of pollution generated.
- *Environmental certainty:* Environmental taxes and charges increase the cost of products and activities in a direct and predictable way, making it easier to judge the associated positive environmental impacts.[7,8]

Implementation of Environmental Taxes and Charges

Environmental taxes and charges are commonly used on certain products such as packaging to reflect the costs of end-of-life disposal. A common scheme is deposit-refund systems in which regulators require a monetary deposit at the time of a product being sold. The deposit is eventually refunded when the item is returned for recycling or correctly disposed of.[9,10,11] An alternative to taxing environmental 'bads' is to provide tax relief for environmental 'goods'. The tax system can be used to subsidise environment-friendly goods, services, or actions, for example, tax breaks can be used to support research and development (R&D) projects related to environmental technologies.[12,13]

Case: The State of New York's Waste Tire Management Fee

The State of New York's Environmental Conservation Law has imposed a waste tyre and recycling fee on sales of most new tyres sold at retail stores within the state. The New York State Department of Taxation and Finance is responsible for administering the collection of the fee, which has been set at $2.50 per tyre. Tyre retailers, including car dealers and auto repair shops, must collect the fee on most new tyres sold. They must remit $2.25 for each tyre sold when they file the Waste Tire Management Fee return. The remaining $0.25 is kept by the retailer to cover their administrative costs. The fee applies to all new tyres for use on nearly all vehicles including cars, trucks, buses, and motorcycles.[14]

Case: Croatia's Charge on Hazardous Waste

Croatia's Environmental Protection and Energy Efficiency Fund is the central point for collecting and investing extra-budgetary resources in the programmes and projects of environmental and natural protection, energy efficiency, and the use of renewable energy sources. The Fund's Charge on Hazardous Waste is levied on legal and natural persons pursuing a business activity which generates hazardous waste. The charge has been set at HRK 100 per tonne of generated but untreated or non-exported hazardous waste. The amount of the charge on hazardous waste is calculated using the formula

$$N = N_1 \times P \times k_k$$

where:

- N is the amount of the charge on hazardous waste in HRK
- N_1 is the charge per tonne of generated but untreated or non-exported hazardous waste (unit charge)
- P is the quantity of generated but untreated or non-exported hazardous waste in each calendar year
- k_k is the corrective coefficient depending on the properties of hazardous waste[15]

Subsidies and Incentives

Subsidies and incentives, such as grants and low-interest loans, are used to stimulate development of new technologies, help create new markets for environmental goods and services, encourage changes in consumer behaviour through green purchasing schemes, and temporarily support achieving higher levels of environmental protection by companies.[16,17,18,19]

Case: Malaysia's Green Technology Financing Scheme

Malaysia's Green Technology Financing Scheme (GTFS) was introduced by the government to support the development of green technology that:

- minimises the degradation of the environment;
- reduces greenhouse gas emissions;
- is safe for use and promotes a healthy and improved environment for all forms of life;
- conserves the use of energy and natural resources; and
- promotes the use of renewable resources.

GTFS provides soft loans to producers and users in the four key sectors of energy, water and waste management, building, and transport. The loan is supported by the government with the interest rate determined by the participating financial institution: the government bears 2 percent of the total interest/profit charged by the financial institution. The maximum financing amount available to producer companies is RM 100 million, while a maximum amount of RM 10 million is available to user companies with the length of the loan set at 15 years for producers and 10 years for users.[20,21]

Tradeable Permits

Tradeable permits are rights to sell and buy actual or potential pollution in artificially created markets.[22] They are flexible market-based tools to address environmental problems; however, unlike taxes, they are quantity-based, rather than price-based environmental policy measures. The permits can be denominated either in terms of 'bads' (i.e. pollution, emissions) or 'goods' (i.e. ecosystems enhanced).[23,24] There is a quantitative limit on the number of permits available which is either introduced as a maximum ceiling for 'cap and trade' schemes (the cap is a pollution limit with parties exceeding it penalised and the trade part is a market for companies to buy and sell allowances that permit them to emit a certain amount)[25] or as a minimum performance commitment for 'baseline and credit' schemes (the scheme identifies, measures, and provides incentives for activities that reduce emissions or pollution).[26,27]

To ensure that tradeable permit schemes achieve their environmental objectives, two types of monitoring data are required. First, periodic data on the condition of the resource are needed to evaluate the effectiveness of the programme over time. These data are used as the basis for adjusting environmental limits as conditions warrant. Second, resource managers need sufficient data to monitor compliance with the scheme. Monitoring compliance requires data on the identity of permit holders, amounts of permits owned by each holder, and permit transfers. Where programmes have additional restrictions on permit use, for example, types of equipment used or who quotas can be transferred to, the data must be complete enough to contain this information and to identify non-compliance in a timely manner.[28]

Case: Korea's Emissions Trading Scheme

The Korean Emissions Trading Scheme (K-ETS) commenced in 2015 with the aim of Korea reducing its emissions by 37 percent below its business-as-usual emission levels by 2030. In Phase I (2015–2017), 100 percent of the allowances are allocated for free, and in Phase II (2018–2020), 97 percent of the allowances will be allocated for free. Companies whose total annual emissions are 125,000 tonnes of carbon dioxide-equivalent (tCO^2e) or more or companies with places of business whose annual emissions are 25,000 tCO^2e or more are subject to the caps under the K-ETS. The scheme has three types of credits:

- *Korea Allowance Units (KAUs)*: KAUs are allowances allocated to a company subject to targets under the K-ETS.
- *Korea Offset Credit (KOC)*: KOCs are converted from Certified Emission Reductions issued by the Clean Development Mechanism (CDM) for emission reductions achieved by CDM projects under the Kyoto Protocol or other offsets approved by the Korean government. KOCs can be traded between both emissions trading scheme (ETS) and non-ETS entities but cannot be traded in the Korea Exchange. KOCs cannot be submitted to the government for compliance with K-ETS targets.

- *Korea Credit Unit (KCU)*: KCUs are credits converted from KOCs and can be submitted to the government for compliance with K-ETS targets. KOCs can be claimed and traded between ETS entities only and can be traded in the Korea Exchange.

Covered entities can use offsets to meet up to 10 percent of their allowance obligations. In Phase I and Phase II, only domestic offsets are accepted. In the planned Phase III, the use of international offset credits will be allowed, but only up to half of the offsets submitted may be international offset credits. During Phase I, the average closing price of KAUs traded in the Korea Exchange between January 2015 and June 2016 was KRW 16,520/tCO^{2}e with prices reaching 21,000/tCO^{2}e. During Phase I and Phase II, companies, other than those subject to caps under the K-ETS, except for Korea Development Bank, Korea Exim Bank, and the Industrial Bank for Korea, are not allowed to open allowance trading accounts on the K-ETS.[29]

Non-Fiscal Tools

Governments at all levels can use a variety of non-fiscal tools to develop the circular economy and encourage a life cycle perspective to be taken by economic actors in an attempt to decouple economic growth from resource use and associated environmental impacts.

Regulations

Direct regulatory tools, for example, laws or regulations that stipulate environmental quality standards, technology standards, mandatory best practices, or limits to emissions from various pollutants, represent a major proportion of all tools being used for environmental policy. The appeal of using regulations to set standards is that rather than dictate the specific technique for reducing pollution (or improving resource efficiency), regulations, such as setting performance standards, grant firms flexibility in choosing how to meet the standard. Overall, regulations can be used to address a broad range of environmental problems and can make environmental outcomes more certain. They can also be more cost-effective than specific technology mandates.[30,31,32,33]

Case: California's Mandatory Commercial Recycling Law

California's Mandatory Commercial Recycling Law focuses on increasing commercial waste diversion as a method to reduce greenhouse gas emissions. It is designed to achieve a reduction in greenhouse gas emissions of five million metric tCO_2e. As part of the requirements, all businesses, including public entities that generate four cubic yards or more of commercial solid waste per week or multi-family residential dwelling of five units or more, shall arrange for recycling services. Businesses can take one of any combination of the following to reuse, recycle, compost, or otherwise divert solid waste from disposal:

- Self-haul
- Subscribe to a hauler
- Arrange for the pickup of recyclable materials
- Subscribe to a recycling service that may include mixed waste processing that yields diversion results comparable to source separation

The overall benefits of the programme identified by CalRecycle include:

- Opportunities for businesses or multi-family complexes to save money
- Creating jobs in California by providing materials for recycling manufacturing facilities
- Reducing greenhouse gas emissions
- Keeping valuable materials out of landfills
- Creating a healthy environment for the community and future generations by recovering natural resources[34]

Green Public Procurement

Developing green public procurement policies is an effective way of increasing the credibility of public authorities who are encouraging industry and consumers to change their patterns of production and consumption. Green public procurement policies can also be used as a tool for enlarging the market for sustainable products and services.[35,36] Table 2.1 summarises a series of best practices in designing a successful green public procurement policy.[37]

Table 2.1 Best practices in designing green public procurement policies

Best practices	Description
Leadership and commitment from senior managers and policymakers	The impact of green public procurement policies can be enhanced if it has high-level political agreement and is incorporated as part of a broader sustainability policy
Setting and agreeing on sustainability priorities	Targets such as reducing greenhouse gas emissions and priority expenditures on resource efficiency actions optimise the allocation of resources and enable structured decision-making
Mandatory green procurement requirements	This emphasises the priority of the government and provides clear expectations to politicians and procurement officials
Public expenditure management frameworks	These may need adapting to better support procurement policies
Joint procurement by public administration authorities	This can increase bargaining power and help reduce prices associated with relatively more expensive green technologies and products as well as administrative costs per contract
Procurement tools are used to provide guidance to decision-making	These tools may include guidelines, procedures, life cycle assessments, and evaluation impacts
Early engagement with the private sector and other stakeholders	This helps identify the scope for innovation and determine the extent to which local suppliers can respond to stricter standards

Case: Hong Kong's Green Procurement

Since 2000, the Hong Kong Government has required bureaus and departments to consider environmental considerations when procuring goods and services. Specifically, bureaus and departments are encouraged to avoid single-use items and purchase products:

- with improved recyclability, high recycled content, reduced packaging, and greater durability;
- with greater energy efficiency;
- which utilise clean technology and/or clean fuels;
- which result in reduced water consumption;
- which emit fewer irritating or toxic substances during installation or use; and
- which result in the smaller production of toxic substances, or of less toxic substance, upon disposal.

For the purchase of common user items, the government has adopted mandatory specifications when the items are available on the market with adequate ranges of models and quantities of supply. For items with uncertain market availability, the green specifications are included in the tender specifications as 'desirable'. Tenderers are invited to indicate in their offer whether these items can comply with these green features and where appropriate submit supporting documents for verification. The tender assessment panel will then evaluate tender offers. Where there are two or more offers, which are identical in all aspects, the one that can meet the desirable green specifications will be given preference, as summarised in Table 2.2.[38]

Table 2.2 Hong Kong's green procurement assessment

	Product A	Product B	Product C
Price $	$$$	$	$
Mandatory green specification	✓	✓	✓
Desirable green specification	X	✓	X
Tender accepted	No	Yes	No

Enhancing Business Competitiveness

Many governments are recognising the importance of sound environmental management as a competitive advantage and as a driver of economic development, both in terms of new business opportunities and as a spur to innovation. As such, governments have a role in supporting industry sectors and businesses through the promotion of practices that enable them to trade based on sustainable and efficient business practices. This support can come from a variety of services including the following:

- Marketing intelligence about the environmental goods and services sector
- Highlighting opportunities where the environmental goods and services sector might be able to provide solutions to overcome environmental problems
- Helping with engagement and consultation of environmental goods and services stakeholders, trade associations, development agencies, and related clusters[39,40]

Case: Vancouver's Green and Digital Demonstration Program

Vancouver's Green and Digital Demonstration Program (GDDP) provides businesses and start-ups with access to City of Vancouver resources including buildings, streets, vehicles, and digital infrastructure for product testing and showcasing of opportunities. GDDP is jointly delivered by the Vancouver Economic Commission (VEC) and the City of Vancouver, with VEC and the city lending staff time to support the launch and implementation of pilot projects. VEC collects and screens applications, guides successful candidates through the demonstration process, and promotes successful projects. Meanwhile, the city is responsible for providing support and staff time to manage the installation and operation of the pilot on available assets, which could be physical installations on city assets or the implementation and testing of digital programme by city staff. With only a limited number of places available on the programme, applications are assessed on the following:

- *Scalability*: Company and technology can be scaled up, generate future sales, and create local jobs.
- *Environmental impact (for green applicants)*: A direct and measurable impact on the environment is projected.
- *Implementability*: Products and companies are near market-ready or are market-ready and fully compatible with an available city asset.
- *Minimal risk*: Technology has been vetted by a third-party stakeholder.
- *No direct cost*: No direct or incremental costs shall be incurred by the city for implementation.[41,42]

The overall benefits of participating in the programme include:

- Refining solutions
- Accelerating the commercialisation of a company's technology
- Attracting investment
- Gaining traction and sales in the marketplace
- Leveraging the City of Vancouver's $31 billion green and innovative brand[43]

Cluster Policies

Many governments have initiated eco-industrial cluster policies that enable firms to benefit from industrial symbiosis, exchange by-products among companies, and form inter-firm networking arrangements that increase material productivity and reduce operating costs. The firms involved in eco-industrial clusters are typically from new industries,

Table 2.3 Unintended benefits of cluster partnerships

Benefit	Description
Attitude change	Greater understanding and valuing of other sectors/communities
Networking	The development of new and trusted connections
Human capital	Improved working practices and human capital development from exposure to different working methods and viewpoints
Social capital	An improvement in reputation and hence an increased willingness of others to work with or trust an organisation
Spin-off partnerships	An increased interest, capacity, and opportunity to build future successful partnerships such as partnerships to commercialise R&D

rather than traditional manufacturing sectors. Overall, there are four major sources of productivity gains and positive cost-benefits that can be linked to eco-industrial clusters:

1. The effective use of raw and waste materials
2. Access to knowledge and technology
3. Employment generation
4. Complementary eco-product development[44]

Nevertheless, while the objectives of the eco-clusters may be clearly defined, there are often a range of unintended benefits attributed to the partnerships formed, summarised in Table 2.3.[45,46]

Case: Italy's National Technology Cluster of Green Chemistry

In 2012, Italy identified eight permanent technology clusters with the aim of creating closer ties between industry, universities, and regional and national institutions. The National Technology Cluster of Green Chemistry—Sustainable Processes and Resources for Innovation and National Growth (SPRING) aims to trigger the growth and development of bio-based industries in Italy through the revitalising of Italian chemistry in the name of environmental, social, and economic sustainability. The areas SPRING focuses on include:

- *Renewable resources as raw material*: Determination of the most appropriate local species (scraps or dedicated crops) to use in biorefinery processes

- *Biorefineries*: Creation of third-generation biorefineries in local areas to obtain high-value products such as biochemicals and biomaterials
- *Bio-based products*: Development and promotion of new bio-based products (partially or totally obtained from renewable resources) with low environmental impact
- *Support R&D activities*: Support activities that contribute to the growth of investments in innovative technologies and in pilot plants to stimulate the bioeconomy at a regional and national level[47,48,49]

Education and Training

The transition towards a circular economy requires a requisite skills base of not only trained professionals with technical, management, and business knowledge, but of society too. Education and training should be a continuous task across the educational curriculum that endorses lifelong learning about both the environmental gains and economic opportunities associated with the circular economy. There are different stages of education and training in which resource efficiency can be integrated including the following:

- *Primary and secondary school*: Primary education plays a role in shaping mindsets towards environmental protection and resource efficiency. During this stage, individuals establish a value set that enables them to make informed decisions in the future that can increase quality of life while protecting the environment and promoting resource efficiency.
- *Technical and vocational*: This training provides knowledge, methods, and tools to help students design and implement practical solutions on resource efficiency as they enter the workforce.
- *Higher education*: Universities prepare future decision-makers of society for their entry into the labour market. Disciplines relevant to resource efficiency and environmental protection include engineering, management, economics, and public administration.
- *Lifelong/on the job*: Education and training for resource efficiency is a continuous process that can be offered to employees and society.[50]

Case: Singapore's Professional Sharing Series

The Singapore Environment Institute's (SEI) Professional Sharing Series (PSS) aims to promote a dynamic exchange of environmental knowledge between the National Environment Agency and industries, academics, and businesses. PSS is a knowledge-sharing platform where SEI invites professionals in their respective fields to share their knowledge, expertise, and perspectives on environmental issues. Topics covered include emerging environmental technologies, environmental management approaches, and good environmental practices.[51]

Raising Industry Awareness and Capacity

Persuading industries to improve their environmental performance is often a challenging task; for instance, industries and businesses, including small and medium-sized enterprises (SMEs), need to maintain their competitiveness and therefore are reluctant to adopt resource efficiency measures that they perceive will increase costs. At other times, businesses are unaware of resource efficiency technologies and practices that exist. As such, disseminating and demonstrating the benefits of environmentally sustainable and resource-efficient practices and technologies is an effective way of fostering the adoption of new production methods. To do so, governments can choose to work directly with individual companies or in partnerships with industrial associations. Activities that can result in immediate environmental and economic benefits for businesses include:

- providing energy- and raw material-saving advice;
- making available funds to small resource efficiency investments; and
- providing resources to monitor and disseminate the results of circular economy-related projects.[52,53]

Case: Sweden's Coaches for Energy and Climate

Sweden's Coaches for Energy and Climate programme provides free coaching for SMEs (that have an annual energy usage below 300,000 kilowatt hours (kWh)) to reduce their energy usage. Companies that join the programme will get an on-site visit from the coach followed by a review of the company's energy usage. The company will then be provided with suggestions on areas that have the greatest potential for reducing energy usage. In addition, the company will be provided individual coaching for implementing specific energy efficiency measures. Throughout the programme, the coach will maintain regular contact with the company and support them in implementing the energy efficiency measures. During the programme, companies will be invited to meet other companies involved in the coaching programme. At the meetings, there will be an opportunity to learn more about energy efficiency in general and share experiences with others.[54]

Case: Invest Northern Ireland's Sustainable Development Support Programme

Invest Northern Ireland's Sustainable Development Support Programme provides up to five days of free consultancy support to help businesses increase resource efficiency and lower waste. The support is available to companies with a total resource spend of over GBP 30,000 a year. Consultancy support can help businesses in the following ways:

- Developing robust business cases for individual projects
- Providing project management skills on projects
- Managing the purchase and installation of new equipment
- Defining equipment and/or process specifications
- Identifying suppliers of goods and services
- Encouraging collaborative efforts with other businesses in the area around energy[55]

Industry-Based Standards

Industry-based standards help increase resource efficiency progress by defining and promoting standards for products made by firms and by facilitating quality control to help firms comply with environmental standards. Standards play a key role in the design, manufacturing, packaging, and end-of-life stages of products and services. Standardisation also helps the acceptance of innovations by markets. Industry-based standards are commonly integrated into voluntary agreements or environmental management systems. There are four types of approaches, in increasing order of the importance, that government authorities play in their application:

1. *Unilateral commitments made by polluters*: This consists of environmental improvement programmes set up by firms and communicated to their stakeholders. The definition of environmental targets, as well as

the provisions governing compliance, is determined by the firms themselves. Nonetheless, firms may delegate monitoring and dispute resolution to a third party to strengthen the agreement's credibility and environmental effectiveness.

2. *Private agreements between polluters and polluted*: This involves contracts being formed between a firm or group of firms and those that are being harmed by their emissions. The contract stipulates the undertaking of an environmental management programme and/or setting of a pollution abatement device.

3. *Environmental agreements negotiated between industry and public authorities*: This involves contracts being formed between public authorities (local, national, federal, or regional) and industry. They often contain a target (e.g. a pollution abatement objective) and a time schedule to achieve it. The public authority commitment generally consists of not introducing a new piece of legislation, for example, a compulsory environmental standard or an environmental tax, unless a voluntary action fails to meet the agreed target.

4. *Voluntary agreements designed by public authorities, to which individual firms are invited to participate in*: Within this type of agreement, participating firms agree to standards (relating to performance, technology, or management) which have been developed by public bodies such as environmental agencies. The agreement defines the conditions of individual membership, the provisions to be complied with by the firms, the monitoring criteria, and the evaluation of the results. The economic benefits for companies being part of the agreement include access to R&D subsidies and technical assistance and enhanced reputation, for example, being permitted to use the associated environmental logo on their products and services.[56,57]

Voluntary Agreements

Voluntary agreements involving government parties encourage businesses, industries, or sectors to improve their resource efficiency and environmental performance beyond regulatory measures. Voluntary agreements range from initiatives where participating firms set the targets and conduct their own monitoring and reporting to initiatives where a contract is made between the private party and a public body. Voluntary agreements can be useful for several reasons including the following:

- Raising awareness and getting buy-in from businesses and the industry on the need for action
- Offering more ambitious goals than regulations while lowering administrative and enforcement costs
- Shifting businesses' mindsets from reactionary attitudes to more proactive and innovative behaviours[58]

To facilitate the success of voluntary agreements, the Organisation for Economic Co-operation and Development (OECD) has detailed a set of recommendations that are summarised in Table 2.4.[59]

Table 2.4 Recommendations on the design of voluntary agreements

Recommendation	Description
Clearly defined targets	The targets should be transparent and clearly defined. Agreements should define quantitative targets. In addition, interim objectives should be listed so parties can identify difficulties of implementation at an early stage
Characterisation of a business-as-usual scenario	Before setting the targets, estimates of a business-as-usual trend should be given to provide a baseline for comparison
Credible threats	At the negotiating stage, a threat provides companies with incentives to go beyond the business-as-usual trend
Credible and reliable monitoring	Monitoring and reporting are essential for tracking performance improvements and avoiding failure to reach the targets
Third-party participation	Involving third parties in the process of setting up agreement objectives and monitoring of performance increases the agreement's credibility. Environmental performance should be made public and transparent to provide additional incentives to respect the commitment
Penalties for non-compliance	Sanctions for non-complying firms should be set
Information-orientated provisions	To maximise the informational soft effects of agreements, support for activities in technical assistance, technical workshops, publication of best practice guides, and so on, should be promoted
Provisions reducing the risk for competition distortions	In the case of collective agreements, safeguards against adverse effects on competition could be provided by notifying anti-trust authorities of new agreements

Case: Finland's Voluntary Energy Efficiency Agreement Scheme

Finland has a new Voluntary Energy Efficiency Agreement scheme that runs between 2017 and 2025. The scheme is a tool chosen together by the government and industrial/municipal associations to fulfil the EU energy efficiency obligations set for Finland. By ensuring the scheme is comprehensive and successful, Finland can continue to meet its obligations without resorting to separate new legislation or other new coercive measures. At the same time, the scheme creates green growth and open markets for clean technology solutions. There will be four Energy Efficiency Agreements signed under the scheme that will cover the following sectors:

1. *Industries (industry, energy sector, and private service sector)*: Companies join the agreement in their own industrial branch (energy, food and drink, chemical, technology, wood, energy production, energy services, motor trade, repairs, and commerce) and commit themselves to improving their energy efficiency in accordance with the actions and targets presented in their respective industrial branch's plan.
2. *Property sector*: Companies sign a separate Accession Document into the Action Plan of Rental Housing or Commercial Properties and commit themselves to improving their energy efficiency.
3. *Municipal sector*: This agreement between the national government and municipalities is to improve the efficient use of energy in the municipal sector. Municipalities, cities, and joint municipalities sign their own Energy Efficiency Agreement and commit themselves to the actions and targets specified in the Energy Efficiency Agreement for municipal sector.
4. *Oil sector (distribution of liquid heating fuels)*: This agreement is between the Ministry of Employment and the Economy, the Ministry of the Environment, the Finnish Petroleum and Biofuels Association, and the major distributors of liquid heating fuels.

When joining the appropriate agreement, the company, industry, or municipality sets an indicative target for quantitative energy savings (MWh) for the entire period 2017–2025 and an intermediate target for 2020. The target for 2025 equals 7.5 percent and the target for 2020 equals 4 percent of the participant's current annual energy use. Participants are required to undertake continuous improvements in energy efficiency and are committed to the following:

- Organising and planning the measures
- Clarifying the possibilities for improvement in energy efficiency
- Implementing cost-effective improvement measures in energy efficiency
- Taking energy efficiency into consideration in planning and purchasing
- Training the staff and communicating on energy efficiency matters
- Annual reporting
- Striving to implement new energy-efficient technology and increase the use of renewable energy sources[60]

Eco-Labels and Certification

Eco-labels and certification encourage sustainable consumption by providing consumers with information about the environmental impact of products and services. Companies are also rewarded for designing products and services that have the least environmental impact, therefore, further encouraging attitudes towards more pro-environmental management. Eco-labels and certifications can apply to the entire life cycle of a product or a specific stage or step in the production process. They are generally awarded by a third party that authorises the use of the label/certification on products in a certain product category.[61]

> **Case: New Zealand's Environmental Choice Label**
>
> Environmental Choice is New Zealand's official environmental label, initiated and endorsed by the New Zealand Government but independently operated. The label identifies products and services that have a lower impact on the environment, have an ongoing commitment to the New Zealand environment, and contribute towards a more sustainable economy. The label can be trusted by consumers as it applies a rigorous multi-criteria assessment, looks at the entire life cycle of the product, includes an independent third-party certification process, and is subject to an independent peer review to ensure compliance with the International Organization for Standardization (ISO) 14020/24 principles. To ensure continued compliance, all products and services that carry the Environmental Choice label are required to complete an annual audit.[62]

Supporting Life Cycle Analyses

Government policy has tended to focus largely on point sources of pollution such as industrial emissions rather than the products themselves and how they contribute to environmental degradation. The concept of a life cycle analysis is to minimise environmental impact by looking at all phases of a product's life cycle and acting where it is most effective. The development of policies to support life cycle analyses can be guided by four principles:

1. Pollution and waste reduction measures need to be identified throughout the product's entire life cycle.
2. Policy measures need to be flexible, working alongside the market where possible.

3. The actions of stakeholders need to be recognised as the environmental impacts of products is determined by the actions of many, including designers, manufacturers, marketing people, retailers, and consumers.
4. Instead of setting final targets to be reached, the principle of continuous product improvement needs to be stimulated.[63]

Case: Life Cycle Kerbside Recycling Calculator in Victoria, Australia

The State of Victoria in Australia has developed an interactive Life Cycle Kerbside Recycling Calculator for local government, businesses, and households to understand the importance of recycling, not just in the use of materials but also in other environmental areas. The 'kerbside' version of the calculator requires large users to enter material streams in tonnages while the 'household' version requires users to enter materials used on a weekly basis. Both versions calculate the annual environmental benefit derived from recycling the products, including greenhouse gas reductions and water and energy savings.[64]

Greening the Supply Chain

Companies are often encouraged or mandated by governments to ensure environmental standards are maintained throughout the entire supply chain. In this way, buyer companies can ensure that the environmental standards that they have adopted internally are consistently maintained by their suppliers. The benefits to both suppliers and their customers go beyond cost-cutting; it can create business value in the form of higher quality materials or manufacturing processes, innovative new goods and services, protection of one's brand, and enhanced customer loyalty.[65]

Case: US EPA's Improving Environmental Sustainability in Supply Chains Webinar

The US EPA (Environmental Protection Agency) held an Improving Environmental Sustainability in Supply Chains: Best Practices Webinar. The webinar involved three leading companies highlighting the importance of engaging and rewarding suppliers to drive sustainability, ways to align and clarify values for suppliers, ways to incentivise suppliers to meet key environmental and health and safety criteria, and ways to build internal organisational support for engaging suppliers on sustainability.[66]

Environmental Recognition Awards

Environmental recognition awards can help raise environmental awareness of businesses, the community, and individuals and help companies gain recognition for their environmental performance. To be effective, the awards need to be widely promoted, for example, in business and industry media. Environmental recognition awards can also be used to recognise the role of different stakeholders including individuals as well as entrepreneurs and small businesses developing circular economy concepts.[67,68]

Case: Calgary's Environmental Achievement Award

Calgary's Environmental Achievement Award recognises environmental contributions made by individuals or organisations, including Calgary-based businesses and corporations, that reduce impacts on and/or restore the city's natural environment. The award recognises innovative environmental contributions including the application of advanced green technology as well as environmental management initiatives that conserve natural resources, reduce pollution (air, land, and water), and promote environmental stewardship.[69]

Extended Producer Responsibility

Extended Producer Responsibility (EPR) is where the producer's responsibility for a product is extended to the post-consumer stage of a product's life cycle. EPRs can be implemented in a variety of ways including take-back programmes, minimum recycled content standards, energy efficiency standards, disposal bans and restrictions, advance disposal fees, virgin materials taxes, and deposit/refund schemes. When implementing EPRs, it is recommended that it is established in accordance with general good governance principles that are summarised in Table 2.5.[70,71,72]

Table 2.5 Extended producer responsibility good governance principles

Principle	Description
Clearly defined objectives based on analysis and consultation with all relevant stakeholders	EPRs usually aim to achieve one or more of four main goals: (1) reducing the use of resources and materials; (2) waste prevention; (3) reducing the environmental impacts of products; and (4) closing material loops. Each EPR should clearly specify which of these goals it aims to achieve
Ensure consistency and coherence with related policies	A life cycle approach ensures that environmental impacts are not increased or transferred somewhere else in the product chain
The scope of the EPR should be clearly defined	Products with the greatest environmental impacts should be the main target; however, a range of other factors influence which product/waste to focus on and how the EPR should be designed, including the durability and composition of the product, the primary and secondary markets in which they are traded, and their distribution networks and supply chain
The producers of the product subject to the EPR should be clearly defined	The producer is the entity with the greatest control over the selection of materials and the design of the product
A consultation process should be organised when establishing an EPR system	The process should aim to enhance its acceptability, transparency, and effectiveness. Subsequently, a communication strategy should be developed to keep all stakeholders informed of the EPR's operations
Flexibility in the phase-in stage	Specific challenges may arise in the start-up phase such as uncertainty about waste volumes and the need for large capital investments in collection and treatment facilities. As such, consideration should be given about specific measures that may be required to help the phase-in of the EPR

Case: Canada's Extended Producer Responsibility Regulations

In Canada, where EPR regulations exist, all manufacturers, retailers, distributors, and other suppliers of certain products must participate in approved product stewardship plans. For example, Call2Recycle Canada, Inc. has

approved product stewardship plans for household/consumer batteries in British Columbia, Manitoba, and Quebec. Under this scheme, parties are obligated to participate in the scheme if they answer yes to one or more of the following questions under the respective province they are in:
- *British Columbia*: Is the party a manufacturer, seller/distributor, or importer?
- *Manitoba*: Is the party a manufacturer or seller/distributor?
- *Quebec*: Is the party an enterprise, brand owner, or importer?[73]

Knowledge Transfer Networks

Governments can facilitate circular economy linkages between enterprises, institutions, universities, private research labs, SMEs, consultants, and entrepreneurs through knowledge networks. These networks are important for the development, nurturing, and dispersing of environmental technologies and practices.[74]

Case: Urbantech NYC

The New York City Economic Development Corporation's Urbantech NYC programme offers entrepreneurs office space, equipment, and a variety of resources so they can address some of New York City's most pressing urban challenges in sectors including energy, waste, transportation, food, water, and the built environment (summarised in Table 2.6).[75] Specifically, Urbantech NYC will provide:

- over 100,000 square feet of affordable and flexible co-working and private office space for companies across all stages of the growth cycle;
- co-located prototype and testing equipment for rapid on-site product development;
- tenant support services such as business skills workshops, mentoring programmes, and organised 'demo days';
- targeted sector-specific events to strengthen the cleantech and smart city communities in New York City;
- workforce development programmes that will train the local workforce and create a talent pipeline for member companies;
- access to a top-tier network of academic institutions, investors, corporate planners, and city agencies.[76]

Table 2.6 Urbantech NYC's sectoral focus

Sector	Description
Energy	Producing, storing, delivering, or consuming energy more cleanly and efficiently
Mobility	Improving how people and goods move within and between cities
Waste	Minimise generating of waste, diverting waste from landfills, improving collection efficiency, and material recovery
Built environment	Making buildings more efficient, sustainable, resilient, connected, and liveable
Food systems	Improving food access, promoting local food production, reducing environmental impact, and reducing food waste
Water	Ensuring clean drinking water, sustainable processing of wastewater, and reducing consumption
Digital	Connect and monitor the physical urban infrastructure to improve planning/investment decisions, help maintain equipment, and measure performance in real-time

Information-Based Tools

Information-based tools enable users to identify environmental challenges and make informed consumption and production decisions.[77] A common type of information-based tool developed by governments are resource maps that contain data on existing and projected resource consumption, sources of surpluses, current networks, and potential networks, and barriers and opportunities. These types of maps can be used to identify opportunities for investment, facilitate stakeholder engagement, and raise awareness of ongoing projects and their benefits.[78]

Case: Brussels' Greencheck

Brussels' Greencheck enables electricity customers to check the part of their electricity supply that is green. To use the tool, customers enter their electricity code, from which they will receive a percentage of green supply. To prove the origin of the green electricity that suppliers sell, suppliers are obliged to do the following:

- Communicate monthly to the distribution and/or transmission system operator the list of customers supplied with green electricity. They must

also indicate to each customer the share of green electricity in the total supply of electricity.
- For each period, submit warranty labels of origin for the green electricity they have supplied during the period.

BRUGEL then approves the proportion of green electricity from the fuel mix of the suppliers only with the help of the warranty labels of origin, which are issued per MWh of electricity produced.[79]

Best Practice Measures for Selecting Effective Tool Mixes

No single tool alone can effectively promote the creation of a circular economy. Instead, an optimal mix of tools is required to achieve enhanced environmental outcomes along with economic objectives.[80] A series of best practice measures for effectively selecting a mix of tools is summarised in Table 2.7.[81,82]

Table 2.7 Best practice measures for selecting effective tool mixes

Best practice measure	Description
Flexibility	Tool mixes should be constructed in an economically efficient manner that provides flexibility and does not stifle innovation
Combining instruments	Combining information-based tools with measures that more directly target the environmental externality can make both instruments more effective
Avoiding tool overlap	Some types of tools that overlap can hamper the proper working of the tools involved or cause unnecessary administration costs
Broad application	Tools should be applied as broadly as possible. For multi-aspect environmental challenges, tools that address total amounts of pollution should supplement tools that address the way a certain product is used, when it is used, and where it is used.
Avoid transferring problems	It is important to avoid transferring environmental problems between different sectors
Policy design	It is important to formulate tools to ensure that they are preventative rather than focusing on 'end-of-pipe' issues

Notes

1. EEA. 2005. Market-based instruments for environmental policy in Europe. Available: https://www.cbd.int/financial/doc/eu-several.pdf.
2. Stavins, R. 2002. Experience with market-based environmental policy instruments. Available: https://www.econstor.eu/bitstream/10419/119660/1/NDL2002-052.pdf.
3. Brears, R. C. 2017. *The Green Economy and the Water-Energy-Food Nexus.* London, Palgrave Macmillan UK.
4. Eurostat. 2017. *Environmental taxes* [Online]. Available: http://ec.europa.eu/eurostat/web/environment/environmental-taxes [Accessed 11 August 2017].
5. Japan Center for a Sustainable Environment and Society. 2017. *What are environmental taxes?* [Online]. Available: http://www.jacses.org/en/paco/envtax.htm.
6. OECD. 2011. Environmental taxation: A guide for policy makers. Available: https://www.oecd.org/env/tools-evaluation/48164926.pdf.
7. Fullerton, D., Leicester, A. & Smith, S. 2008. Environmental taxes. Cambridge, MA: National bureau of economic research.
8. OECD. 2011. Environmental taxation: A guide for policy makers. Available: https://www.oecd.org/env/tools-evaluation/48164926.pdf.
9. OECD. 2016b. Policy guidance on resource efficiency. Available: http://www.oecd.org/environment/waste/Resource-Efficiency-G7-2016-Policy-Highlights-web.pdf.
10. EEA. 2005. Market-based instruments for environmental policy in Europe. Available: https://www.cbd.int/financial/doc/eu-several.pdf.
11. World Bank. 2012b. Market-based instruments / economic incentives. Available: http://siteresources.worldbank.org/INTRANETENVIRONMENT/Resources/GuidanceNoteonMarketBasedInstruments.pdf.
12. OECD. 2011. Environmental taxation: A guide for policy makers. Available: https://www.oecd.org/env/tools-evaluation/48164926.pdf.
13. OECD. 2016b. Policy guidance on resource efficiency. Available: http://www.oecd.org/environment/waste/Resource-Efficiency-G7-2016-Policy-Highlights-web.pdf.
14. New York State Department of Taxation and Finance. 2017. *Waste tire management fee* [Online]. Available: https://www.tax.ny.gov/bus/tire/wtm.htm [Accessed 11 August 2017].
15. Energy Efficiency and Environmental Protection Fund. 2017. *Charge on hazardous waste* [Online]. Available: http://www.fzoeu.hr/en/environ-

mental_fees/fees_pursuant_to_the_act_on_the_environmental_protection_and_energy_efficiency_fund/charges_on_burdening_the_environment_with_waste/charge_on_hazardous_waste/ [Accessed 11 August 2017].

16. OECD. 2012. OECD environmental outlook to 2030. Available: http://www.oecd-ilibrary.org/environment/oecd-environmental-outlook-to-2030_9789264040519-en.

17. World Bank. 2012a. Guidance notes on tools for pollution management. Available: http://siteresources.worldbank.org/INTRANETEN-VIRONMENT/Resources/GuidanceNoteonMarketBasedInstruments.pdf.

18. UNIDO. 2011. UNIDO green industry policies for supporting green industry. Available: https://www.unido.org/fileadmin/user_media/Services/Green_Industry/web_policies_green_industry.pdf.

19. Resources for the Future. 2007. Environmental and technology policies for climate mitigation.

20. GreenTech Malaysia. 2017b. *Green technology financing scheme. Empowering green businesses* [Online]. Available: https://www.gtfs.my/.

21. GreenTech Malaysia. 2017a. *Frequently asked questions* [Online]. Available: https://www.gtfs.my/faq#n24130.

22. OECD. 2017. *Tradable pollution permits* [Online]. Available: https://stats.oecd.org/glossary/detail.asp?ID=2737.

23. OECD. 2004. Tradeable permits. Policy evaluation, design and reform. Available: http://www.oecd-ilibrary.org/environment/tradeable-permits_9789264015036-en.

24. OECD. 2012. OECD Environmental Outlook to 2030. Available: http://www.oecd-ilibrary.org/environment/oecd-environmental-outlook-to-2030_9789264040519-en.

25. EDF. 2017. *How cap and trade works* [Online]. Available: https://www.edf.org/climate/how-cap-and-trade-works.

26. Australian Goverment Climate Change Authority. 2014. Coverage, additionality and baselines – Lessons from the Carbon Farming Initiative and other schemes. Available: http://climatechangeauthority.gov.au/reviews/coverage-additionality-and-baselines-lessons-carbon-farming-initiative-and-other-schemes.

27. OECD. 2012. OECD environmental outlook to 2030. Available: http://www.oecd-ilibrary.org/environment/oecd-environmental-outlook-to-2030_9789264040519-en.

28. Tietenberg, T. 2003. Tradable permits in principle and practice. Available: http://web.mit.edu/ckolstad/www/TT_SBW.pdf.

29. IETA. 2016. Republic of Korea: An emissions trading case study. Available: http://www.ieta.org/resources/2016%20Case%20Studies/Korean_Case_Study_2016.pdf.
30. Goulder, L. H. & Parry, I. W. H. 2008. Instrument choice in environmental policy. *Review of Environmental Economics and Policy*, 2, 152–174.
31. OECD. 2008a. An OECD framework for effective and efficient environmental policies. Available: https://www.oecd.org/env/tools-evaluation/41644480.pdf.
32. OECD. 2012. OECD environmental outlook to 2030. Available: http://www.oecd-ilibrary.org/environment/oecd-environmental-outlook-to-2030_9789264040519-en.
33. Goulder, L. H. & Parry, I. W. H. 2008. Instrument choice in environmental policy. *Review of Environmental Economics and Policy*, 2, 152–174.
34. CalRecycle. 2017. *Mandatory commercial recycling* [Online]. Available: http://www.calrecycle.ca.gov/recycle/commercial/.
35. Witjes, S. & Lozano, R. 2016. Towards a more Circular Economy: Proposing a framework linking sustainable public procurement and sustainable business models. *Resources, Conservation and Recycling*, 112, 37–44.
36. UNIDO. 2011. UNIDO green industry policies for supporting green industry. Available: https://www.unido.org/fileadmin/user_media/Services/Green_Industry/web_policies_green_industry.pdf.
37. Rainville, A. 2016. Standards in green public procurement – A framework to enhance innovation. *Journal of Cleaner Production*.
38. Environmental Protection Department. 2017. *Green procurement* [Online]. Available: http://www.epd.gov.hk/epd/english/how_help/green_procure/green_procure.html.
39. UNIDO. 2011. UNIDO green industry policies for supporting green industry. Available: https://www.unido.org/fileadmin/user_media/Services/Green_Industry/web_policies_green_industry.pdf.
40. OECD. 2015. Environmental policy toolkit for greening SMEs in the EU eastern partnership countries. Available: https://www.oecd.org/environment/outreach/Greening-SMEs-policy-manual-eng.pdf.
41. Vancouver Economic Commission. 2017b. *Green and digital demonstration program* [Online]. Available: http://www.vancouvereconomic.com/gddp/.

42. Vancouver Economic Commission. 2017a. *GDDP – Acceptance criteria* [Online]. Available: http://www.vancouvereconomic.com/gddp/gddp-acceptance-criteria/.

43. Vancouver Economic Commission. 2017b. *Green and digital demonstration program* [Online]. Available: http://www.vancouvereconomic.com/gddp/.

44. UNIDO. 2011. UNIDO green industry policies for supporting green industry. Available: https://www.unido.org/fileadmin/user_media/Services/Green_Industry/web_policies_green_industry.pdf.

45. OECD. 2016b. Policy guidance on resource efficiency. Available: http://www.oecd.org/environment/waste/Resource-Efficiency-G7-2016-Policy-Highlights-web.pdf.

46. OECD. 2010. Cluster policies. *OECD Innovation Policy Platform* [Online]. Available: http://www.oecd.org/innovation/policyplatform/48137710.pdf.

47. SPRING. 2017. *Home* [Online]. Available: http://www.clusterspring.it/home-en/.

48. ALISEI. 2017. *National technological clusters* [Online]. Available: http://www.clusteralisei.it/en/national-technological-clusters/.

49. Research Italy. 2017. *Green chemistry* [Online]. Available: https://www.researchitaly.it/en/national-technology-clusters/green-chemistry/#null.

50. UNIDO. 2011. UNIDO green industry policies for supporting green industry. Available: https://www.unido.org/fileadmin/user_media/Services/Green_Industry/web_policies_green_industry.pdf.

51. National Environment Agency. 2016. PSS 68: Making do with less – Improving resource efficiency. Available: http://www.nea.gov.sg/docs/default-source/training-knowledge-hub/sei/pss68_edm.pdf.

52. UNIDO. 2011. UNIDO green industry policies for supporting green industry. Available: https://www.unido.org/fileadmin/user_media/Services/Green_Industry/web_policies_green_industry.pdf.

53. OECD. 2015. Environmental policy toolkit for greening SMEs in the EU eastern partnership countries. Available: https://www.oecd.org/environment/outreach/Greening-SMEs-policy-manual-eng.pdf.

54. Swedish Energy Agency. 2017. *Coaches for energy and climate* [Online]. Available: https://energimyndigheten.a-w2m.se/Home.mvc.

55. Invest NI. 2017. *Manage business energy and waste* [Online]. Available: https://www.investni.com/support-for-business/manage-business-energy-and-waste.html.

56. UNIDO. 2011. UNIDO green industry policies for supporting green industry. Available: https://www.unido.org/fileadmin/user_media/Services/Green_Industry/web_policies_green_industry.pdf.
57. OECD. 2008b. Voluntary approaches for environmental policy. Effectiveness, efficiency and usage in policy mixes. Available: http://www.oecd-ilibrary.org/environment/voluntary-approaches-for-environmental-policy_9789264101784-en.
58. UNIDO. 2011. UNIDO green industry policies for supporting green industry. Available: https://www.unido.org/fileadmin/user_media/Services/Green_Industry/web_policies_green_industry.pdf.
59. OECD. 2008b. Voluntary approaches for environmental policy. Effectiveness, efficiency and usage in policy mixes. Available: http://www.oecd-ilibrary.org/environment/voluntary-approaches-for-environmental-policy_9789264101784-en.
60. Motiva. 2017. *Energy efficiency agreements* [Online]. Available: http://www.energiatehokkuussopimukset2017-2025.fi/en/energy-efficiency-agreements/.
61. UNIDO. 2011. UNIDO green industry policies for supporting green industry. Available: https://www.unido.org/fileadmin/user_media/Services/Green_Industry/web_policies_green_industry.pdf.
62. Environmental Choice New Zealand. 2017. *Products and services* [Online]. Available: https://www.environmentalchoice.org.nz/products-and-services/.
63. UNIDO. 2011. UNIDO green industry policies for supporting green industry. Available: https://www.unido.org/fileadmin/user_media/Services/Green_Industry/web_policies_green_industry.pdf.
64. Sustainability Victoria. 2017. *Life cycle assessment of kerbside recycling in Victoria* [Online]. Available: http://www.sustainability.vic.gov.au/publications-and-research/research/life-cycle-assessment.
65. UNIDO. 2011. UNIDO green industry policies for supporting green industry. Available: https://www.unido.org/fileadmin/user_media/Services/Green_Industry/web_policies_green_industry.pdf.
66. U.S. EPA. 2017. *Improving environmental sustainability in supply chains: Best practices webinar* [Online]. Available: https://www.epa.gov/climateleadership/improving-environmental-sustainability-supply-chains-best-practices-webinar.
67. OECD. 2015. Environmental policy toolkit for greening SMEs in the EU eastern partnership countries. Available: https://www.oecd.org/environment/outreach/Greening-SMEs-policy-manual-eng.pdf.

68. Brears, R. C. 2017. *The green economy and the water-energy-food nexus.* London, Palgrave Macmillan UK.

69. City of Calgary. 2017. *The environmental achievement award* [Online]. Available: http://www.calgary.ca/CA/city-clerks/Pages/Administration-services/Calgary-Awards/Environmental-Achievement-Awards.aspx.

70. OECD. 2016a. Extended producer responsibility. Updated guidance for efficient waste management. Available: http://www.oecd.org/environment/waste/extended-producer-responsibility-9789264256385-en.htm.

71. UNIDO. 2011. UNIDO green industry policies for supporting green industry. Available: https://www.unido.org/fileadmin/user_media/Services/Green_Industry/web_policies_green_industry.pdf.

72. OECD and Ministry of the Environment Japan. 2014. The state of play on extended producer responsibility (ERP) opportunities and challenges. *Global forum on environment: Promoting sustainable materials management through extended producer responsibility (EPR).* Tokyo.

73. Call2Recycle. 2017. *Am I obligated?* [Online]. Available: http://www.call2recycle.ca/am-i-obligated/.

74. UNIDO. 2011. UNIDO green industry policies for supporting green industry. Available: https://www.unido.org/fileadmin/user_media/Services/Green_Industry/web_policies_green_industry.pdf.

75. URBANTECHNYC. 2017. *About* [Online]. Available: http://www.urbantechnyc.com/about/.

76. NYCEDC. 2017. *Urbantech NYC* [Online]. Available: https://www.nycedc.com/program/urban-technology-growth-hubs.

77. OECD. 2012. OECD environmental outlook to 2030. Available: http://www.oecd-ilibrary.org/environment/oecd-environmental-outlook-to-2030_9789264040519-en.

78. Brears, R. C. 2017. *The green economy and the water-energy-food nexus.* London, Palgrave Macmillan UK.

79. BRUGEL. 2017. *Provision of green electricity in the Brussels-Capital Region* [Online]. Available: http://www.brugel.be/fr/secteur-de-l-energie/electricite-verte-et-energie-renouvelable/fourniture-de-electricite-verte-en-region-de-bruxelles-capitale.

80. UNIDO. 2011. UNIDO green industry policies for supporting green industry. Available: https://www.unido.org/fileadmin/user_media/Services/Green_Industry/web_policies_green_industry.pdf.

81. Ibid.

82. OECD. 2012. OECD environmental outlook to 2030. Available: http://www.oecd-ilibrary.org/environment/oecd-environmental-outlook-to-2030_9789264040519-en.

References

ALISEI. 2017. *National technological clusters* [Online]. Available: http://www.clusteralisei.it/en/national-technological-clusters/.

Australian Goverment Climate Change Authority. 2014. Coverage, additionality and baselines – Lessons from the Carbon Farming Initiative and other schemes. Available: http://climatechangeauthority.gov.au/reviews/coverage-additionality-and-baselines-lessons-carbon-farming-initiative-and-other-schemes.

Brears, R.C. 2017. *The green economy and the water-energy-food nexus*. London: Palgrave Macmillan UK.

BRUGEL. 2017. *Provision of green electricity in the Brussels-capital region* [Online]. Available: http://www.brugel.be/fr/secteur-de-l-energie/electricite-verte-et-energie-renouvelable/fourniture-d-electricite-verte-en-region-de-bruxelles-capitale.

Call2Recycle. 2017. *Am I obligated?* [Online]. Available: http://www.call2recycle.ca/am-i-obligated/

CalRecycle. 2017. *Mandatory commercial recycling* [Online]. Available: http://www.calrecycle.ca.gov/recycle/commercial/.

City of Calgary. 2017. *The environmental achievement award* [Online]. Available: http://www.calgary.ca/CA/city-clerks/Pages/Administration-services/Calgary-Awards/Environmental-Achievement-Awards.aspx

EDF. 2017. *How cap and trade works* [Online]. Available: https://www.edf.org/climate/how-cap-and-trade-works.

EEA. 2005. Market-based instruments for environmental policy in Europe. Available: https://www.cbd.int/financial/doc/eu-several.pdf.

Energy Efficiency and Environmental Protection Fund. 2017. *Charge on hazardous waste* [Online]. Available: http://www.fzoeu.hr/en/environmental_fees/fees_pursuant_to_the_act_on_the_environmental_protection_and_energy_efficiency_fund/charges_on_burdening_the_environment_with_waste/charge_on_hazardous_waste/. Accessed 11 Aug 2017.

Environmental Choice New Zealand. 2017. *Products and services* [Online]. Available: https://www.environmentalchoice.org.nz/products-and-services/

Environmental Protection Department. 2017. *Green procurement* [Online]. Available: http://www.epd.gov.hk/epd/english/how_help/green_procure/ green_procure.html.

Eurostat. 2017. *Environmental taxes* [Online]. Available: http://ec.europa.eu/ eurostat/web/environment/environmental-taxes. Accessed 11 Aug 2017.

Fullerton, D., A. Leicester, and S. Smith. 2008. *Environmental taxes.* Cambridge, MA: National Bureau of Economic Research.

Goulder, L.H., and I.W.H. Parry. 2008. Instrument choice in environmental policy. *Review of Environmental Economics and Policy* 2: 152–174.

GreenTech Malaysia. 2017a. *Frequently asked questions* [Online]. Available: https://www.gtfs.my/faq#n24130

———. 2017b. *Green technology financing scheme. Empowering green businesses* [Online]. Available: https://www.gtfs.my/.

IETA. 2016. Republic of Korea: An emissions trading case study. Available: http://www.ieta.org/resources/2016%20Case%20Studies/Korean_Case_ Study_2016.pdf.

Invest NI. 2017. *Manage business energy and waste* [Online]. Available: https:// www.investni.com/support-for-business/manage-business-energy-and-waste. html.

Japan Center for a Sustainable Environment and Society. 2017. *What are environmental taxes?* [Online]. Available: http://www.jacses.org/en/paco/envtax. htm.

Motiva. 2017. *Energy efficiency agreements* [Online]. Available: http://www. energiatehokkuussopimukset2017-2025.fi/en/energy-efficiency-agreements/.

National Environment Agency. 2016. PSS 68: Making do with less – Improving resource efficiency. Available: http://www.nea.gov.sg/docs/default-source/ training-knowledge-hub/sei/pss68_edm.pdf.

New York State Department of Taxation and Finance. 2017. *Waste tire management fee* [Online]. Available: https://www.tax.ny.gov/bus/tire/wtm.htm. Accessed 11 Aug 2017.

NYCEDC. 2017. *Urbantech NYC* [Online]. Available: https://www.nycedc. com/program/urban-technology-growth-hubs.

OECD. 2004. Tradeable permits. Policy evaluation, design and reform. Available: http://www.oecd-ilibrary.org/environment/tradeable-permits_ 9789264015036-en.

———. 2008a. An OECD framework for effective and efficient environmental policies. Available: https://www.oecd.org/env/tools-evaluation/41644480.pdf.

————. 2008b. Voluntary approaches for environmental policy. Effectiveness, efficiency and usage in policy mixes. Available: http://www.oecd-ilibrary.org/environment/voluntary-approaches-for-environmental-policy_9789264101784-en.

————. 2010. Cluster policies. *OECD Innovation Policy Platform* [Online]. Available: http://www.oecd.org/innovation/policyplatform/48137710.pdf.

————. 2011. Environmental taxation: A guide for policy makers. Available: https://www.oecd.org/env/tools-evaluation/48164926.pdf.

————. 2012. OECD environmental outlook to 2030. Available: http://www.oecd-ilibrary.org/environment/oecd-environmental-outlook-to-2030_9789264040519-en.

————. 2015. Environmental policy toolkit for greening SMEs in the EU eastern partnership countries. Available: https://www.oecd.org/environment/outreach/Greening-SMEs-policy-manual-eng.pdf.

————. 2016a. Extended producer responsibility. Updated guidance for efficient waste management. Available: http://www.oecd.org/environment/waste/extended-producer-responsibility-9789264256385-en.htm.

————. 2016b. Policy guidance on resource efficiency. Available: http://www.oecd.org/environment/waste/Resource-Efficiency-G7-2016-Policy-Highlights-web.pdf.

————. 2017. *Tradable pollution permits* [Online]. Available: https://stats.oecd.org/glossary/detail.asp?ID=2737.

OECD and Ministry of the Environment Japan. 2014. The state of play on extended producer responsibility (ERP) opportunities and challenges. *Global forum on environment: Promoting sustainable materials management through extended producer responsibility (EPR)*. Tokyo.

Rainville, A. 2016. Standards in green public procurement – A framework to enhance innovation. *Journal of Cleaner Production* 167: 1029–1037.

Research Italy. 2017. *Green chemistry* [Online]. Available: https://www.researchitaly.it/en/national-technology-clusters/green-chemistry/#null.

Resources for the Future. 2007. Environmental and technology policies for climate mitigation.

SPRING. 2017. *Home* [Online]. Available: http://www.clusterspring.it/home-en/.

Stavins, R. 2002. Experience with market-based environmental policy instruments. Available: https://www.econstor.eu/bitstream/10419/119660/1/NDL2002-052.pdf.

Sustainability Victoria. 2017. *Life cycle assessment of kerbside recycling in Victoria* [Online]. Available: http://www.sustainability.vic.gov.au/publications-and-research/research/life-cycle-assessment.

Swedish Energy Agency. 2017. *Coaches for energy and climate* [Online]. Available: https://energimyndigheten.a-w2m.se/Home.mvc.

Tietenberg, T. 2003. Tradable permits in principle and practice. Available: http://web.mit.edu/ckolstad/www/TT_SBW.pdf.

U.S. EPA. 2017. *Improving environmental sustainability in supply chains: Best practices webinar* [Online]. Available: https://www.epa.gov/climateleadership/improving-environmental-sustainability-supply-chains-best-practices-webinar.

UNIDO. 2011. UNIDO green industry policies for supporting green industry. Available: https://www.unido.org/fileadmin/user_media/Services/Green_Industry/web_policies_green_industry.pdf.

Urbantech NYC. 2017. *About* [Online]. Available: http://www.urbantechnyc.com/about/.

Vancouver Economic Commission. 2017a. *GDDP – Acceptance criteria* [Online]. Available: http://www.vancouvereconomic.com/gddp/gddp-acceptance-criteria/.

———. 2017b. *Green and digital demonstration program* [Online]. Available: http://www.vancouvereconomic.com/gddp/.

Witjes, S., and R. Lozano. 2016. Towards a more Circular Economy: Proposing a framework linking sustainable public procurement and sustainable business models. *Resources, Conservation and Recycling* 112: 37–44.

World Bank. 2012a. Guidance notes on tools for pollution management. Available: http://siteresources.worldbank.org/INTRANETENVIRONMENT/Resources/GuidanceNoteonMarketBasedInstruments.pdf.

———. 2012b. Market-based instruments/economic incentives. Available: http://siteresources.worldbank.org/INTRANETENVIRONMENT/Resources/GuidanceNoteonMarketBasedInstruments.pdf.

3

Natural Resource Management and the Circular Economy in London

Introduction

London's economy is specialised in a variety of professional, scientific, and technical services including finance, insurance, information technology, and communication. Employment in these industries accounted for over 70 percent of all jobs in 2014.[1] London's gross value added (GVA) growth rate is forecasted to increase from 2.9 percent in 2016 to 3.4 percent in 2017 and 3.3 percent in 2018. This corresponds to a projected rise in employment and household income and spending over the same time frame.[2]

Challenges to the Linear Economy

London is experiencing a variety of challenges to its traditional linear economy as described below through an assortment of examples.

© The Author(s) 2018
R. C. Brears, *Natural Resource Management and the Circular Economy*,
Palgrave Studies in Natural Resource Management,
https://doi.org/10.1007/978-3-319-71888-0_3

Air Pollution

In 2008, there were an estimated 4300 deaths in London due to long-term exposure to small particles. In 2015, a research study found that in 2010 there was an equivalent of 5900 deaths across the city associated with NO2. There is still today a significant exposure of the population to levels of NO2 above the European Union (EU) limit value and while this exposure is predicted to decline significantly by 2020 (−96 percent), current modelling shows that in 2020 there will still be more than 72,000 people living in locations with average NO2 levels above the EU limit value. This contrasts with average concentrations of particles (PM10 and PM2.5) already being within EU limit value by 2013.[3]

Climate Change

In London, summers will become hotter with the average summer day in 2050 projected to be 2.7°C warmer and very hot days 6.5°C higher than the baseline average of 1961–1990. By 2100, the average summer day is likely to be 3.9°C warmer and the hottest day of the year could be 10°C hotter than the hottest day this decade. By 2050, summers will be drier with the average summer expected to be 19 percent drier and the driest summer 39 percent drier than the baseline average. By the end of the century, average summers could be 23 percent drier. Meanwhile, in 2050, the average winter will likely be 15 percent wetter and the wettest winter 33 percent wetter than the baseline average.[4]

Energy

Much of London's energy demand (around 94 percent) is sourced from outside the city. Due to limited space, London could never be self-sufficient in energy even if energy demand is reduced and more renewable energy is generated within the city boundaries. Regarding carbon emissions, London's electricity demand accounts for almost half of the city's total carbon dioxide emissions. Existing climate change and energy programmes in the city have achieved a reduction of 670 kilotonnes of carbon dioxide equivalent (ktCO2e) in 2015, a threefold increase over 2010; however, moving through the century, reducing electricity demand and carbon emissions further will be challenged by a growing population and electrification of heat and transport.[5]

Population Growth

Between mid-2011 and mid-2015, London's population increased by 5.7 percent, compared with 2.9 percent for the United Kingdom.[6] By 2050, London's population is expected to reach 11.1 million compared to around 8.7 million today.[7] This makes the population growth in the city more than twice that of Wales, Scotland, Northern Ireland, and the three northern English regions. The population is growing due to almost 200,000 people from overseas arriving in the city each year along with an average of 130,000 births a year.[8]

Waste

London's homes, public buildings, and businesses produce around 7 million tonnes of waste per annum, of which just 52 percent is recycled. At the household level, the waste recycling rate increased from 8 to 30 percent between 2003 and 2010; however, this trend has stalled at 32 percent and remains below the national average of 44 percent. The city's waste bill is around GBP 2 billion a year and is rising. Regarding waste disposal, incineration of waste has doubled from 900,000 tonnes in 2011 to 1.8 million tonnes in 2016, producing around 560,000 tCO2e emissions. Despite the amount of waste collected for landfill falling from 65 percent to 20 percent over the past 10 years, it is estimated that the city's landfill capacity will run out by 2026.[9]

Water

London's average water consumption is 156 litres per person per day, which is around 10 percent higher than the national average of 139 litres per person per day. London's growing population and businesses are demanding more water with the city forecasted to have a water resource gap of over 100 million litres per day by 2020, rising to a deficit of over 400 million litres per day by 2040. At the same time, London is at risk of drought if reservoirs and groundwater aquifers are not refilled by regular rainfall. Thames Water has estimated that the cost of a severe drought to London's economy would be GBP 330 million per year.[10]

Upstream Fiscal Tools

London has implemented a variety of upstream fiscal tools to develop the circular economy and encourage a life cycle perspective to be taken by economic actors in an attempt to decouple economic growth from resource use and associated environmental impacts.

Flexible Financing for Circular Economy Business Projects

Over the period 2015–2020, London Waste and Recycling Board (LWARB) is allocating GBP 20 million towards circular economy business projects. The funding is flexible and can be tailored to suit the requirements of each project, with the funding limit set by state aid legislation and LWARB's own risk appetite. The investment products available include project finance, corporate loans, growth equity, and fund investments. In addition, LWARB will explore ways to support start-ups and early-stage ventures developing innovative circular economy business models or products. The funding will be distributed on a first-come, first-served basis with eligibility based on

- A strong linkage to London's economy, such as being headquartered in the city or having a substantial proportion of operations, supply chain, or customers located within London
- A robust, scalable, and deliverable business plan
- An experienced, high-quality management team
- An existing funding gap[11]

Overall, the types of projects likely to be funded include the following business models:

- *Those using renewable inputs*: Shifting to the use of renewable energy or secondary materials in the input process
- *Those that recover value*: Recovering value at the end of the product life through biological or technical recycling
- *Those that prolong product life*: Done through smart design, use of data analytics, remanufacturing, and maintenance
- *Those operating in the sharing economy*: Sharing of assets including cars, rooms, and machinery
- *Those selling products as services*: Selling access to products while retaining ownership or dematerialising products, for example, books or online shopping[12]

Upstream Non-Fiscal Tools

London has implemented a variety of upstream non-fiscal tools to develop the circular economy and encourage a life cycle perspective to be taken by economic actors in an attempt to decouple economic growth from resource use and associated environmental impacts.

London Leaders Programme

The London Leaders programme, run by the London Sustainable Development Commission with support of the Mayor of London, identifies and nurtures some of London's innovative new leaders in sustainable businesses and communities by supporting them in demonstrating sustainability in action and their contribution to the London economy. Specifically, the programme supports innovation and green entrepreneurship on projects that will boost London's economy while supporting the principles of sustainable development. Each London Leader receives support if required in the form of

- Training and master classes
- Specialist mentoring

- Profile building and recognition as a London Leader
- Links with relevant networks[13]

Business Energy Challenge

The Business Energy Challenge recognises businesses' efforts to reduce energy use. In 2015, businesses competed for a Gold, Silver, or Bronze award and/or one of the new special awards. The main award is based on the percentage carbon intensity reduction per m^2 for a business's London property portfolios (inclusive of owned, leased, whole/part of building) from the baseline (2010–2011) to the challenge year (2014–2015). The percentage reductions of entrants were ranked according to the greatest reduction with the businesses falling into one of the bands described in Table 3.1. The percentage reduction in carbon intensity per m^2 is based on annual energy use converted to carbon emitted and divided by the total floor space for that year. This created a value for the intensity of carbon emitted by each company. The percentage reduction in carbon intensity allowed large and small companies to be compared more fairly. Special awards were also distributed for businesses leading the way in reducing carbon emissions (summarised in Table 3.2).

Clean City Awards Scheme

Since 1994, London has run the Clean City Awards Scheme to encourage partnerships with all types of city businesses by raising the profile of responsible waste management and recognising and rewarding good

Table 3.1 Business energy challenge awards

Award	Qualifying boundaries
Gold	Top 10 percent of businesses making the greatest percentage reduction in carbon intensity per m^2
Silver	Following 15 percent
Bronze	Following 20 percent
Recognition of participation	All other businesses

Mayor of London. 2015. *About the 2015 awards* [Online]. Available: https://www.london.gov.uk/what-we-do/environment/energy/about-2015-awards

Table 3.2 Business energy challenge special business awards

Awards	Description
Sector Leader (automatic entry)	The sector award was based on the same performance metric as the main award, but businesses were only compared with their peers from the same sector, allowing recognition of sector-specific achievements. Sectors included finance and insurance, manufacturing, retail units, accommodation, food service, and entertainment
Novel Climber of the Challenge (automatic entry)	Businesses that were new to the Business Energy Challenge were entered for the best newcomer award. The performance was based on the same metric as the main award, but they were only compared to other businesses entering the Business Energy Challenge for the first time
Climber of the Year (additional data may be required)	Businesses that entered 2013/2014 were eligible for the award with the business making the largest reduction over the one-year period the winner
Large Portfolio Climber of the Challenge (automatic entry)	For businesses that entered over 30 locations, the same metric for performance was used as the main award, but businesses were only compared to other businesses that entered over 30 locations
Climber of the Decade (additional data required)	Businesses could choose to enter additional data to make 2005/2006 the baseline year to be eligible for this award
Courageous Climber	This award was for all participants that opted in and allowed the Business Energy Challenge to publish their non-anonymous data in the public domain
Team Climb	This award was for the best landlord/tenant/managing agent/facility manager or energy manager collaboration. To enter, businesses had to report the actions that at least two parties listed above had taken since 2010/2011 to improve energy efficiency and reduce carbon and the buildings that these actions related to. The award was judged by a panel appointed by the Business Energy Challenge

Mayor of London. 2015. *About the 2015 awards* [Online]. Available: https://www.london.gov.uk/what-we-do/environment/energy/about-2015-awards

practice by encouraging businesses to 'reduce, reuse, and recycle'. The Awards Scheme is open to all businesses based in the City of London who are members of the scheme with each business coming under one of three membership categories: Small Sites, Large Sites, and Facilities Managers. Each year members can apply for an award to recognise their efforts and commitment to adopting sustainable waste management practices. After an initial form-based assessment, city officers carry out a site inspection to assess the waste management practices in place and gather further information on waste minimisation, reuse, and recycling initiatives. The officers look for documentary evidence such as waste transfer notes to confirm initiatives are in place and that the applicant is meeting legal requirements. Each site is then scored to establish which award—merit, gold, gold commendation, and platinum—should be presented. The highest scoring sites in each category—Large Sites, Small Sites, and Facilities Managers—are invited to present to a panel of judges to potentially win the 'Chairman's Cup' in their category, which is known as Final Judging. Those invited to Final Judging give a short presentation detailing the initiatives in place, achievements to date in the year, and the reasons why they deserve to win the Chairman's Cup. This is followed by questions from the panel. In addition to awarding the Chairman's Cup for each category, the judges also award special commendations for outstanding efforts worthy of recognition. Overall, the Awards Scheme aims to

- Promote good waste management practices
- Encourage waste minimisation, reuse, and recycling
- Reduce the amount of waste sent to landfill
- Ensure compliance with Duty of Care regulations
- Encourage businesses to take pride in their surroundings
- Provide a forum for businesses to exchange waste management initiatives
- Reduce smoking-related litter[14]

Advance London: Circular Economy SME Business Support Programme for London

The Advance London programme provides free practical help and advice to small and medium-sized enterprises (SMEs) in London to enable businesses to adopt and scale up circular economy business models. Specifically, a team of highly skilled business advisors will work with businesses to identify the opportunities available and provide practical support, including market analysis, option appraisals, financial modelling, business case development, and facilitating access to investment and funding, to pilot and implement circular economy business models to achieve a variety of benefits including

- Increasing their competitiveness
- Accessing new markets
- Growing and expanding
- Improving resource efficiency
- Driving innovation
- Building supply chain linkages
- Becoming more resilient to external factors[15]

Circular Economy Champions Programme

The Circular Economy Champions Programme involves LWARB working with three people per year, each of whom will actively promote circular economy thinking and practices within their organisations. LWARB offers each champion a place at the Ellen MacArthur Foundation's Executive Education course and a fully funded place at the foundation's circular economy acceleration workshop or annual summit. As part of the programme, each champion is required to actively promote circular economy thinking and practices within their organisation and networks with LWARB arranging three workshops per year for the champions to attend, where practical steps to accelerate the circular economy are discussed.[16]

The Mayor's Entrepreneur

The Mayor's Entrepreneur is a competition that invites London students to come up with smart ideas that improve the city. Students submit their idea under the categories of air quality, water, energy, transport, food waste, recycling and reuse, and 'other' with one overall winner receiving a prize of GBP 20,000. Each submission is judged on

- *Originality*: What makes the idea new for London? How is it better and smarter than what is already out there?
- *Practicality*: Is the top prize of GBP 20,000 enough to make the idea a reality?
- *Clarity*: How clear is the idea?
- *Longevity*: What does the idea look like in five years' time?
- *Carbon savings*: If the idea is smarter than other businesses, then it's saving energy and resources, which means saving carbon too.[17]

Sustainable Design and Construction Guidance

In 2014, the Mayor of London published supplementary planning guidance on sustainable design and construction which provides guidance to architects, developers, engineers, local planning authorities, and neighbourhoods to achieve sustainable development. The guidance

- Provides detail on how to implement sustainable design and construction
- Provides guidance on how to develop more detailed local policies on sustainable design and construction
- Provides best practice guidance on how to meet the sustainability targets set out in the London Plan
- Provides examples on how to implement sustainability measures within developments

  Some of the areas of guidance include:

- Energy-efficient design
- Meeting carbon dioxide reduction targets
- Urban greening
- Pollution control[18]

Smart London Investor Showcase

In March 2016, the Mayor of London, in partnership with the UK Business Angels Association, held the Smart London Investor Showcase initiative to showcase start-ups and small companies developing 'smart' solutions to tech investors. The entrepreneurs showcased innovative solutions that use data and digital technologies to tackle some of London's challenges as the city continues to grow to over 10 million in 2036. The entrepreneurs who participated needed to have a demonstrable product or service which had passed the proof-of-concept stage and had a physical presence within the city. The entrepreneurs had to tackle one of five challenges that included (1) environment, (2) buildings and homes, (3) transport, (4) health, and (5) resilience and infrastructure. Regarding the environment challenge, the entrepreneurs had to have a product or service that answered the question of how can people interact with, measure, and make the most of London's environment to improve the sustainability and liveability of the city, including

- Reducing emissions and waste
- Improving air quality
- Enhancing resource efficiency
- Improving access to green spaces[19]

RE:FIT London

RE:FIT London aims to help London's non-domestic public buildings and assets become more energy efficient. Ongoing since 2008, the programme reduces carbon emissions and guarantees cost savings for the public sector. The programme helps a range of organisations including London boroughs, National Health Service bodies, central government departments, schools, and other educational establishments to implement retrofit projects. There are two components to the programme:

1. *RE:FIT London Programme Delivery Unit*: RE:FIT's highly skilled and experienced Programme Delivery Unit (PDU) provides free-of-charge

support to public-sector organisations to help them get energy retrofit projects and programmes up and running and successfully implemented. The PDU tailors its advice to each organisation by assessing the potential for retrofitting, formulating retrofit projects, providing funding and procurement advice, and supporting organisations through the procurement process including during and after project delivery. The PDU also provides a range of best practice information, including case studies and benchmarking, and cost information on project costs, resource efficiency savings, and carbon reductions.

2. *RE:FIT Framework for Energy Service Companies*: Most organisations participating in RE:FIT London will use the RE:FIT Framework for Energy Service Companies to procure an energy services company to carry out retrofits because it offers guaranteed energy and cost savings. To help public-sector organisations procure an energy services company quickly, efficiently, and economically, the framework comprises the energy service companies who have an excellent track record in providing energy reduction savings for organisations.[20]

Hydrogen London

Hydrogen London, established by the Mayor of London in 2002, is a partnership consisting of experts from government, business, and academia that aims to bring hydrogen and fuel cell technology to London, secure new jobs, and investment, and support London's resource-efficient economy. Some of the key tasks of Hydrogen London are

- *Dialogue*: Hydrogen London helps create dialogue among key industry stakeholders
- *Funding*: Hydrogen London offers a platform for funding bids
- *Forums*: Hydrogen London helps set up forums to share hydrogen technology research and materials
- *Publications*: Hydrogen London creates reports that outline the benefits of hydrogen fuel cell technologies and steps to develop these further[21]

Downstream Fiscal Tools

London has implemented a variety of downstream fiscal tools to develop the circular economy and encourage a life cycle perspective to be taken by economic actors in an attempt to decouple economic growth from resource use and associated environmental impacts.

Mayor's Air Quality Fund

The Mayor's Air Quality Fund has begun distributing GBP 20 million to support London's boroughs in improving their air quality. In the first round of funding, GBP 5 million was awarded to a wide range of projects including reducing pollution from construction sites. In round two, GBP 5 million was awarded including support for the London Low Emission Construction Partnership (LLECP).

London Low Emission Construction Partnership

The LLECP is a partnership between the 'Cleaner Air Boroughs' of Camden, Hammersmith and Fulham, Islington, Lambeth, Lewisham, and Wandsworth, industry partners across the demolition and construction sector, and King's College London. The objectives of the project are to help the industry understand its impact on local air quality, encourage the uptake of 'best in class' pollution reduction (abatement) measures at construction sites, improve pollution monitoring and make data available for construction sites in London, and evaluate the cost-effectiveness of pollution abatement techniques.[22,23]

Crowdfund London

The Crowdfund London initiative involves people pitching project ideas that respond to local challenges or opportunities in a creative way, help the local economy, and are environmentally sustainable. The winning idea receives a Mayoral pledge of up to GBP 50,000 for the project with the remaining required amount raised through crowdfunding. The City of London is working with the crowdfunding platform Spacehive enabling people to pitch their project, big or small, and potentially win the pledge. To be eligible, the project must be in Greater London; well-resourced with a clear plan and budget; and managed by an organisation representing a local community.[24]

Downstream Non-Fiscal Tools

London has implemented a variety of downstream non-fiscal tools to develop the circular economy and encourage a life cycle perspective to be taken by economic actors in an attempt to decouple economic growth from resource use and associated environmental impacts.

FoodSave

Between 2013 and 2015, the FoodSave project helped small and medium-sized food businesses in London reduce their food waste, put surplus food to good use, and dispose of unavoidable food waste more responsibly through processes including composting or anaerobic digestion. The project worked with around 200 businesses to achieve the programme's goal of

- Diverting over 1000 tonnes of food waste from landfill
- Reducing food waste by over 150 tonnes
- Saving businesses over GBP 350,000 through waste reduction and disposal initiatives

To encourage the wise use of surplus food the Sustainable Restaurant Association (SRA) delivered the FoodSave programme to food service and hospitality businesses including restaurants, pubs, staff canteens, hotels, and cafes. SRA worked with businesses to run 'food waste audits' over a two-week period, which involved SRA implementing a three-step process:

1. Carrying out a free detailed food waste audit of the business
2. Analysing the data to understand where and why waste is being generated
3. Providing practical guidance on how to reduce the waste being produced[25]

The Mayor's Biodiesel Programme

The Mayor's Biodiesel Programme, a partnership between the Mayor of London's Office, the fuel industry, and local authorities, aims to inspire a biodiesel industry revolution with the city capturing, retaining, and processing its used cooking oils (UCO) and fats, oils, and greases (FOG) into biodiesel to power all of London's buses and public-sector road fleet. The benefits are that it will provide a local renewable fuel supply, reduce carbon emissions, and avoid UCO and FOG blockages in the London sewer network.[26]

London Waste Map

The London Waste Map is the city's first attempt to visually present London waste data. The map identifies operational waste sites across the city receiving waste as well as areas identified by boroughs as potential waste sites. Types of waste identified by the map include

- Disposal
- Organic treatment
- Fuel prep, mechanical biological treatment, and thermal treatment
- Household reuse and recycling centres
- Material recycling/sorting
- Metals and vehicle recycling
- Waste transfer (household and commercial)
- Waste transfer (construction)
- Storage
- Licensed tonnage <= 15,000
- Other waste[27]

London Heat Map

The London Heat Map is an online tool that enables users to find opportunities for decentralised energy (DE) projects in London. It contains local information to help users, including local councils and developers, identify and develop DE opportunities and includes data on:

- Major energy consumers
- Fuel consumption and carbon emissions
- Energy supply plants
- Community heating networks
- Heat density[28]

For developers, the London Heat Map provides a dynamic tool that allows users to download and update information and data on the map. Specifically, by registering to access and download data, developers can also add new information on potential heat loads or supply opportunities.[29]

Cleaner Vehicle Checker Scheme

London has initiated the online Cleaner Vehicle Checker Scheme to tackle air pollution in the city. The scheme will show Londoners how much toxic NOx new cars emit, helping them choose and buy less polluting vehicles. Consumers will be able to type in the model of a new car or van they are considering buying and find out more about its actual 'on the road' emissions. This will

- Help people make an informed choice and minimise the number of 'more polluting' vehicles being bought and used in London.
- Help consumers recognise the environmental benefits of switching to zero or ultra-low emission vehicles and encourage their purchase.
- Encourage manufacturers to produce vehicles that conform to the EU's full 'real-world driving emissions' standards much sooner than legally required by 2021.
- Provide a free 'health check' service to London fleet operators to understand how their current fleet performs and where significant improvements could be made.

- Create a tool to allow the Greater London Authority Group and local authorities to lead by example and only buy or lease the cleanest vehicles.[30]

Circular Economy Hackathon

LWARB partnered with RMW and SUEZ Recycling and Recovery UK to host a circular economy hackathon (#CEHACK17) at the RWM exhibition in September 2017 with a GBP 7000 prize for the winning entry. #CEHACK17 aimed to showcase innovative ideas that use mobile technology, smart data, and the Internet of Things to deliver solutions for improving circularity. The hackathon was open to entrepreneurs, start-ups as well as small businesses. Innovations at #CEHACK17 could address, but were not limited to, the following challenges:

- Influence consumers' purchasing decisions by helping them buy products that can be repaired, exchanged, or recycled.
- Develop apps that can help with the collection of small volume or specialist waste or redundant products.
- Inform consumers on how materials can be recycled and whether their local municipal collection services can collect and recycle those materials.
- Engage 18–34-year-olds with recycling, through gamification or reward schemes.[31]

Case Study Summary

London's specialised economy is projected to grow over the next year along with employment and consumption. Nonetheless, the city is experiencing a variety of challenges to its linear economy. Air pollution poses health risks with Londoners exposed to NO2 levels above EU limits. Climate change will see the city experiencing higher average temperatures, drier summers, and higher rainfall events in winters. London's density hampers the city's ability to be self-sufficient in renewable energy, limiting the city's ability to reduce carbon emissions from electricity demand. This is an issue, given that London's rapid population growth will significantly increase demand for electricity and transport. London's recycling rate at the household level has stalled and remains below the national average while the city's waste bill is rising, and landfill space is rapidly declining. Finally, London's per capita water consumption is significantly above the national average and the water supply network is at risk from drought which would have a severe impact on the city's economy.

Upstream

London has implemented a variety of upstream fiscal and non-fiscal tools to develop the circular economy and encourage a life cycle perspective to be taken by economic actors in an attempt to decouple economic growth from resource use and associated environmental impacts. Some of the main tools are as follows:

- LWARB is funding circular economy business projects through a variety of tailored financial instruments including loans and growth and equity fund investments with types of projects receiving funding including ones that use renewable inputs, recover value, and prolong product life.
- The London Leaders programme nurtures green entrepreneurs as they contribute to the city's circular economy. Support comes in a variety of forms including master classes, mentoring, and showcasing them as a London Leader.
- To encourage businesses to reduce energy use, the city initiated the Business Energy Challenge in which businesses were awarded for their efforts in reducing energy use (and carbon emissions) across their property or property portfolio.
- The Circular Economy Champions Programme run by LWARB selects three circular economy champions per year and provides them with education and skills development to actively promote the circular economy within their respective organisations.
- To facilitate partnerships between businesses on circular economy initiatives, London has been running the Clean City Awards Scheme in which participants are recognised and rewarded for 'reduce, reuse, and recycle' good practices.

- The Mayor's Entrepreneur encourages the development of the circular economy within younger generations with students submitting ideas on how to use resources more efficiently, with the winner receiving a cash prize.
- To support SMEs in the circular economy, LWARB's Advance London programme provides free, practical advice to SMEs to help them adopt and scale up circular economy business models with help including the deployment of business advisors.
- The Mayor of London has published guidance on sustainable design and construction for architects, developers, and engineers to achieve sustainable development in the built environment as well as meet sustainability targets in the city's London Plan.
- The Mayor of London held, in partnership with the UK Business Angels Association, the Smart London Investor Showcase initiative which provided a platform for entrepreneurs and start-ups to demonstrate interactive products or services which contribute towards the development of a smart, circular economy in a variety of economic sectors.
- The Hydrogen London initiative is a public-private partnership between the Mayor of London, government agencies, businesses, and academia to develop hydrogen and fuel cell technology in London. The initiative hosts dialogues, offers funding for projects, and holds forums to share hydrogen technology research.
- London's RE:FIT programme provides a range of tools for non-domestic public buildings and assets to reduce their energy usage, including free-of-charge support for organisations to conduct energy retrofits as well as a framework for public organisations to use when procuring energy services companies to ensure they achieve energy and cost savings.

Downstream

London has implemented a variety of downstream fiscal and non-fiscal tools to develop the circular economy and encourage a life cycle perspective to be taken by economic actors in an attempt to decouple economic growth from resource use and associated environmental impacts. Some of the main tools are as follows:

- The Mayor's Air Quality Fund supports projects that reduce pollution at construction sites.
- The LLECP, involving public and private parties, aims to encourage the construction industry to undertake best practice pollution abatement measures on construction sites.
- The Crowdfund London initiative is a competition in which people pitch projects that aim to develop London's local economy in an environmentally sustainable manner. The winner receives a cash prize which includes a Mayoral pledge along with cash raised through a crowdsourcing platform.
- FoodSave provided advice to London businesses on how to reduce their waste and put surplus food to good use. The programme also provided food waste audits for participating businesses.
- The Mayor's Biodiesel Programme involves the Mayor's office working with the fuel industry and local authorities to develop UCO and FOG into biodiesel to power London's buses and public-sector road fleet.
- London's Waste Map visually identifies operational waste sites across the city as well as potential waste sites for waste disposal, organic treatment, material recycling, and sorting.
- The London Heat Map is an online tool for users, including local councils and developers, to find opportunities for DE projects across the city with the map providing data on major energy consumers, community heating networks, and heat density.
- London's online Cleaner Vehicle Checker Scheme helps Londoners choose and buy less polluting vehicles with consumers able to find out the emissions of specific vehicle types and models.
- LWARB's hackathon, open to entrepreneurs, start-ups, as well as small businesses, aimed to showcase innovative ideas that deliver solutions for improving the circularity of the economy with the winner receiving a cash prize.

Overall, London has implemented a variety of upstream and downstream fiscal and non-fiscal tools to develop the circular economy and encourage a life cycle perspective to be taken by economic actors in an attempt to decouple economic growth from resource use and associated environmental impacts. These tools are summarised in Table 3.3.

Table 3.3 London case study summary

Tool	Tool type	Tool title	Description	Upstream/ Downstream
Fiscal	Subsidies and Incentives	Flexible Financing for Circular Economy Business Projects	LWARB funds circular economy business projects through a variety of instruments including loans and investments	Upstream
		Mayor's Air Quality Award	Funding available for projects that improve air quality across the city including projects that reduce pollution at construction sites	Downstream
Non-fiscal	Green Public Procurement	RE:FIT London	Provides a framework for public organisations to use when procuring energy services companies	Upstream
	Enhancing Business Competitiveness	Smart London Investment Showcase	Provided a platform for entrepreneurs and start-ups to demonstrate interactive circular economy products or services	Upstream
	Education and Training	Circular Economy Champions Programme	Champions provided with education and skills development to actively promote the circular economy within their respective organisations	Upstream
	Raising Industry Awareness and Capacity	London Leaders Programme	Supports innovative green entrepreneurship by nurturing new leaders as they contribute to the city's economy	Upstream
		FoodSave	FoodSave provided advice to businesses on how they could reduce their food waste and put surplus food to good use. Waste audits were also done for participating businesses	Downstream
		Advance London	Free, practical advice given to SMEs to help them adopt and scale up circular economy business models	Upstream
		Sustainable Design and Construction	Guidance for architects, developers, and engineers to achieve sustainable development in the built environment	Upstream
	Environmental Recognition Awards	Business Energy Challenge	Businesses were awarded for their efforts in reducing energy use (and carbon emissions) across their property or property portfolio	Upstream
		Clean City Awards Scheme	Businesses of all sizes are recognised and rewarded for 'reduce, reuse, and recycle' good practices	Upstream
		The Mayor's Entrepreneur	Students submit ideas on how to use resources more efficiently with the winner receiving a cash prize	Upstream
		Crowdfund London	A competition in which people pitch projects that respond to London's challenges and help the economy become sustainable	Downstream
		Circular Economy Hackathon	Aimed to showcase innovative circular economy ideas with the winning entry receiving a cash prize	Downstream
	Knowledge Transfer Networks	Hydrogen London	A public-private knowledge network to develop hydrogen and fuel cell technology in London	Upstream
		LLECP	The public-private partnership helps the construction industry understand its impact on local air quality and encourages best practices in pollution abatement	Downstream
		The Mayor's Biodiesel Programme	A partnership between the city, industry, and local authorities to develop biodiesel for London's buses and public-sector road fleet	Downstream
	Information-Based Tools	London Waste Map	The map identifies operational waste sites across the city and potential waste sites for waste disposal, organic treatment, and material recycling	Downstream
		London Heat Map	An online tool for users to find opportunities for DE projects	Downstream
		Vehicle Checker Scheme	Helps Londoners choose less polluting vehicles to purchase	Downstream

Notes

1. GLA Economics. 2017. A description of London's economy. Available: https://www.london.gov.uk/sites/default/files/description-londons-economy-working-paper-85.pdf.
2. Mayor of London. 2016. *London's economic outlook: Spring 2016* [Online]. Available: https://www.london.gov.uk/business-and-economy-publications/londons-economic-outlook-spring-2016.
3. Mayor of London. 2017i. Updated analysis of air pollution exposure in London – Final report. Available: https://www.london.gov.uk/WHAT-WE-DO/environment/environment-publications/updated-analysis-air-pollution-exposure-london-final.
4. Mayor of London. 2011. Managing risks and increasing resilience: our adaptation strategy. Available: https://www.london.gov.uk/WHAT-WE-DO/environment/environment-publications/managing-risks-and-increasing-resilience-our.
5. Mayor of London. 2017b. Draft London Environment Strategy – have your say. Available: https://www.london.gov.uk/what-we-do/environment/draft-london-environment-strategy-have-your-say.
6. Office for National Statistics. 2016. *Population dynamics of UK city regions since mid-2011* [Online]. Available: https://www.ons.gov.uk/peoplepopulationandcommunity/populationandmigration/populationestimates/articles/populationdynamicsofukcityregionssincemid2011/2016-10-11.
7. Mayor of London. 2017b. Draft London Environment Strategy – have your say. Available: https://www.london.gov.uk/what-we-do/environment/draft-london-environment-strategy-have-your-say.
8. The Guardian. 2016. *London population growth rate twice that of UK, official figures show* [Online]. Available: https://www.theguardian.com/uk-news/2016/oct/12/london-population-growth-twice-that-of-uk-official-figures-show.
9. Mayor of London. 2017b. Draft London Environment Strategy – have your say. Available: https://www.london.gov.uk/what-we-do/environment/draft-london-environment-strategy-have-your-say.
10. Ibid.
11. Circular Economy Club. 2016. *London Waste & Recycling Board – UK* [Online]. Available: https://www.circulareconomyclub.com/lwarb_uk/.
12. LWARB. 2016. Establishing London's circular economy. Available: http://www.insidegovernment.co.uk/uploads/2016/09/wayne-hubbard.pdf.

13. London Leaders. 2015. Identifying and nurturing London's most exciting and innovative new leaders in 20 sustainable business and communities. Available: http://www.londonsdc.org.uk/documents/2015London LeadersBrochure_Digital_FINAL.pdf.

14. City of London. 2017b. *Clean City Awards Scheme* [Online]. Available: https://www.cityoflondon.gov.uk/services/environment-and-planning/waste-and-recycling/commercial-waste-and-recycling/clean-city-awards/Pages/default.aspx, City of London. 2017a. *Awards assessment* [Online]. Available: https://www.cityoflondon.gov.uk/services/environment-and-planning/waste-and-recycling/commercial-waste-and-recycling/clean-city-awards/Pages/Awards-assessment.aspx.

15. LWARB. 2017a. *Advance London* [Online]. Available: http://www.lwarb.gov.uk/what-we-do/advance-london/#circular-business-models.

16. LWARB. 2017c. *Circular economy champions programme* [Online]. Available: http://www.lwarb.gov.uk/what-we-do/circular-london/circular-economy-champions-programme/.

17. Mayor of London. 2017e. *The Mayor's entrepreneur* [Online]. Available: https://www.london.gov.uk/what-we-do/environment/smart-london-and-innovation/mayor-entrepreneur-2017#acc-i-42604.

18. Mayor of London. 2014. Sustainable design and construction. Supplementary planning guidance. Available: https://www.london.gov.uk/sites/default/files/gla_migrate_files_destination/Sustainable%20Design%20%26%20Construction%20SPG.pdf.

19. Mayor of London. 2017h. *Smart London Investor Showcase* [Online]. Available: https://www.london.gov.uk/what-we-do/business-and-economy/science-and-technology/smart-london/mayor-and-smart/smart-london-0#acc-i-43572.

20. Mayor of London. 2017g. *RE:FIT* [Online]. Available: https://www.london.gov.uk/what-we-do/environment/energy/energy-buildings/refit.

21. Hydrogen London. 2017. *About us* [Online]. Available: http://www.hydrogenlondon.org/about-hydrogen-london/.

22. Mayor of London. 2017c. *Mayor's air quality fund* [Online]. Available: https://www.london.gov.uk/what-we-do/environment/pollution-and-air-quality/mayors-air-quality-fund.

23. LLECP. 2017. *About* [Online]. Available: http://www.llecp.org.uk/about/about-project.

24. Mayor of London. 2017a. *Crowdfund London: Create. Fund. Launch* [Online]. Available: https://www.london.gov.uk/what-we-do/regeneration/funding-opportunities/crowdfund-london/about#acc-i-47068.

25. FoodSave. 2017. *About* [Online]. Available: http://www.foodsave.org/about/.
26. Mayor of London. 2017d. *The Mayor's biodiesel programme* [Online]. Available: https://www.london.gov.uk/what-we-do/environment/waste-and-recycling/mayors-biodiesel-programme.
27. Mayor of London. 2017j. *Waste map* [Online]. Available: https://maps.london.gov.uk/webmaps/waste/.
28. Mayor of London. 2017k. *What is the London heat map?* [Online]. Available: https://www.london.gov.uk/what-we-do/environment/energy/london-heat-map/what-london-heat-map.
29. Mayor of London. 2017l. *Who does the heat map help?* [Online]. Available: https://www.london.gov.uk/what-we-do/environment/energy/london-heat-map/who-does-heat-map-help#acc-i-43816.
30. Mayor of London. 2017f. *Mayor unveils new cleaner vehicle checker scheme* [Online]. Available: https://www.london.gov.uk/city-hall-blog/mayor-unveils-new-cleaner-vehicle-checker-scheme.
31. LWARB. 2017b. *#CEHACK17* [Online]. Available: http://www.lwarb.gov.uk/what-we-do/advance-london/investment-for-businesses/innovation-challenges/cehack17/.

References

Circular Economy Club. 2016. *London Waste & Recycling Board – UK* [Online]. Available: https://www.circulareconomyclub.com/lwarb_uk/.
City of London. 2017a. *Awards assessment* [Online]. Available: https://www.cityoflondon.gov.uk/services/environment-and-planning/waste-and-recycling/commercial-waste-and-recycling/clean-city-awards/Pages/Awards-assessment.aspx.
———. 2017b. *Clean City Awards Scheme* [Online]. Available: https://www.cityoflondon.gov.uk/services/environment-and-planning/waste-and-recycling/commercial-waste-and-recycling/clean-city-awards/Pages/default.aspx.
FoodSave. 2017. *About* [Online]. Available: http://www.foodsave.org/about/.
GLA Economics. 2017. A description of London's economy. Available: https://www.london.gov.uk/sites/default/files/description-londons-economy-working-paper-85.pdf.
Hydrogen London. 2017. *About us* [Online]. Available: http://www.hydrogen-london.org/about-hydrogen-london/.

LLECP. 2017. *About* [Online]. Available: http://www.llecp.org.uk/about/about-project.

London Leaders. 2015. Identifying and nurturing London's most exciting and innovative new leaders in 20sustainable business and communities. Available: http://www.londonsdc.org.uk/documents/2015LondonLeadersBrochure_Digital_FINAL.pdf.

LWARB. 2016. Establishing London's circular economy. Available: http://www.insidegovernment.co.uk/uploads/2016/09/wayne-hubbard.pdf.

———. 2017a. *Advance London* [Online]. Available: http://www.lwarb.gov.uk/what-we-do/advance-london/#circular-business-models.

———. 2017b. *#CEHACK17* [Online]. Available: http://www.lwarb.gov.uk/what-we-do/advance-london/investment-for-businesses/innovation-challenges/cehack17/.

———. 2017c. *Circular economy champions programme* [Online]. Available: http://www.lwarb.gov.uk/what-we-do/circular-london/circular-economy-champions-programme/.

Mayor of London. 2011. Managing risks and increasing resilience: our adaptation strategy. Available: https://www.london.gov.uk/WHAT-WE-DO/environment/environment-publications/managing-risks-and-increasing-resilience-our.

———. 2014. Sustainable design and construction. Supplementary planning guidance. Available: https://www.london.gov.uk/sites/default/files/gla_migrate_files_destination/Sustainable%20Design%20%26%20Construction%20SPG.pdf.

———. 2015. *About the 2015 awards* [Online]. Available: https://www.london.gov.uk/what-we-do/environment/energy/about-2015-awards.

———. 2016. *London's economic outlook: Spring 2016* [Online]. Available: https://www.london.gov.uk/business-and-economy-publications/londons-economic-outlook-spring-2016.

———. 2017a. *Crowdfund London: Create. Fund. Launch* [Online]. Available: https://www.london.gov.uk/what-we-do/regeneration/funding-opportunities/crowdfund-london/about#acc-i-47068.

———. 2017b. Draft London Environment Strategy – have your say. Available: https://www.london.gov.uk/what-we-do/environment/draft-london-environment-strategy-have-your-say.

———. 2017c. *Mayor's Air Quality Fund* [Online]. Available: https://www.london.gov.uk/what-we-do/environment/pollution-and-air-quality/mayors-air-quality-fund.

————. 2017d. *The Mayor's biodiesel programme* [Online]. Available: https://www.london.gov.uk/what-we-do/environment/waste-and-recycling/mayors-biodiesel-programme.

————. 2017e. *The Mayor's entrepreneur* [Online]. Available: https://www.london.gov.uk/what-we-do/environment/smart-london-and-innovation/mayor-entrepreneur-2017#acc-i-42604.

————. 2017f. *Mayor unveils new cleaner vehicle checker scheme* [Online]. Available: https://www.london.gov.uk/city-hall-blog/mayor-unveils-new-cleaner-vehicle-checker-scheme.

————. 2017g. *RE:FIT* [Online]. Available: https://www.london.gov.uk/what-we-do/environment/energy/energy-buildings/refit.

————. 2017h. *Smart London Investor Showcase* [Online]. Available: https://www.london.gov.uk/what-we-do/business-and-economy/science-and-technology/smart-london/mayor-and-smart/smart-london-0#acc-i-43572.

————. 2017i. Updated analysis of air pollution exposure in London – Final report. Available: https://www.london.gov.uk/WHAT-WE-DO/environment/environment-publications/updated-analysis-air-pollution-exposure-london-final.

————. 2017j. *Waste map* [Online]. Available: https://maps.london.gov.uk/webmaps/waste/.

————. 2017k. *What is the London heat map?* [Online]. Available: https://www.london.gov.uk/what-we-do/environment/energy/london-heat-map/what-london-heat-map.

————. 2017l. *Who does the Heat Map help?* [Online]. Available: https://www.london.gov.uk/what-we-do/environment/energy/london-heat-map/who-does-heat-map-help#acc-i-43816.

Office for National Statistics. 2016. *Population dynamics of UK city regions since mid-2011* [Online]. Available: https://www.ons.gov.uk/peoplepopulationandcommunity/populationandmigration/populationestimates/articles/populationdynamicsofukcityregionssincemid2011/2016-10-11.

The Guardian. 2016. *London population growth rate twice that of UK, official figures show* [Online]. Available: https://www.theguardian.com/uk-news/2016/oct/12/london-population-growth-twice-that-of-uk-official-figures-show.

4

Natural Resource Management and the Circular Economy in Seattle

Introduction

Seattle and the surrounding region's economy is led by its advanced aerospace and information technology (IT) sectors. More than 90 percent of Boeing's planes are built in the region and 650 other aerospace companies are near the city. Seattle and the surrounding areas are leaders in cloud computing and software development with Microsoft and Amazon leading the way there.[1] Seattle and the state's economy is one of the fastest growing in the country with Washington's economy growing by 3.1 percent in 2016.[2] Regarding income levels, Seattle's median household income of $80,000 is well above the national average.[3]

Challenges to the Linear Economy

Seattle is experiencing a variety of challenges to its traditional linear economy as described below through an assortment of examples.

© The Author(s) 2018
R. C. Brears, *Natural Resource Management and the Circular Economy*,
Palgrave Studies in Natural Resource Management,
https://doi.org/10.1007/978-3-319-71888-0_4

Air Pollution

An American Lung Association assessment graded Seattle as being the 26th worst metropolitan area in terms of air quality across the country, with the city receiving a 'C' for short-term particle pollution. The State of Washington's Department of Ecology estimated that around 1100 Washington residents die each year due to small particle pollution. In addition, small particulates contribute to around 1500 non-fatal heart attacks and 400 emergency room visits for asthma and other breathing difficulties. It is estimated that the direct and indirect costs of poor air quality for citizens, businesses, and state healthcare institutions is nearly $190 million per annum.[4,5]

Climate Change

In the Puget Sound region, under a 'business-as-usual' greenhouse gas scenario, the average annual temperature will increase by 5.5°F mid-century. Under a low greenhouse gas scenario, warming is slightly lower (4.2°F on average in 2050). By the 2050s, the Seattle area will likely experience 18 additional days of temperatures above 86°F. The city will also experience drier summers (22 percent less precipitation on average) and wetter fall, winter, and spring seasons (3–11 percent increase in precipitation on average), relative to the 1970–1999 baseline. There will be more extreme weather events with the heaviest (top 1 percent) 24-hour rain event in the region expected to be 22 percent more intense on average by the 2080s under the business-as-usual scenario while the frequency of heavy rain events will increase, occurring seven days per year by the 2080s compared to two days per year historically (1970–1999). Sea level is projected to rise to two feet on average in Seattle by 2100 and today's 100-year storm surge event will become a monthly event by 2060.[6]

Energy

More than 90 percent of Seattle's electricity is generated from hydropower, serving around 1 million people in the greater Seattle area. The amount of electricity generated from hydroelectric dams depends on the mountain snowpack's water storage capacity. With 40 percent of the United States' hydropower generated in the Northwest, lower stream flows will likely reduce hydroelectric supply to a vast number of customers. As such, the city relies on purchasing electricity from other sources including petroleum, natural gas, coal, as well as nuclear and wind.[7,8]

Population Growth

Between 2010 and 2016, Seattle's population increased from 608,000 to 686,000. Over the period 2015–2016, the city's population increased by 21,000, making it one of the fastest-growing big cities in the United States. In the wider King County area, the population is projected to increase from 1.9 million in 2010 to around 2.4 million in 2040.[9,10]

Waste

Seattle's municipal solid waste generation has generally followed economic trends even as population growth has steadily increased in the city. The overall recycle rate declined in the first few years of the 2000s from 40 percent to 38.2 percent and then steadily increased since 2003 to reach 53.7 percent in 2010. It is projected that overall waste generation will increase gradually over the period 2011–2030.[11]

Water

The population in the Seattle Public Utilities' retail service area and areas served by wholesale contracts is projected to increase by 21 percent and 25 percent respectively.[12] Despite a rising population, total water demand is expected to increase gradually from 126 million gallons of water per day (mgd) currently to 142 mgd by 2039 and then decline slightly to 137 mgd through 2060.[13] Nonetheless, climate change will likely place stress on the city's water resources with changes in average annual precipitation in the Northwest likely to vary, with summer precipitation potentially declining by as much as 30 percent. This would likely result in increased competition for water resources from municipal and industrial users as well as hydropower and agricultural irrigation.[14]

Upstream Fiscal Tools

Seattle has implemented a variety of upstream fiscal tools to develop the circular economy and encourage a life cycle perspective to be taken by economic actors in an attempt to decouple economic growth from resource use and associated environmental impacts.

Saving Water Partnership Rebates

The Saving Water Partnership is a group of 19 local water utilities, including Seattle Public Utilities, who collaborate to help their customers save water. The partnership provides numerous rebates for commercial and industrial water users in Seattle including:

- *Toilet and urinal rebates*: $75–$150 rebate when replacing older toilets and urinals with efficient models
- *Refrigeration system rebates*: Rebates up to 50 percent of costs for projects involving space cooling, refrigeration systems, and industrial ice-makers
- *Kitchen equipment rebates*: Up to $1500 for replacing inefficient commercial kitchen equipment with water-efficient equipment
- *Medical/scientific equipment rebates*: Rebates up to 50 percent for steam sterilisers, medical air and vacuum systems, x-ray processing, and other medical equipment
- *Laundry equipment rebates*: $200 rebates for efficient coin-operated machines or up to 50 percent of large system improvements
- *Other water-saving technologies rebates*: Rebates of up to 50 percent of project costs
- *Cooling tower improvements*: This pilot programme is providing financial incentives for a range of items including electronic water level controllers (not to exceed $3000)[15]

Energy Efficiency Rebates for Industrial Customers

Seattle City Light offers an Energy Smart Service to its small to medium-business customers in which the utility provides an energy survey, pays for an energy analysis/study, and provides a rebate of up to 70 percent of approved energy efficiency upgrades. Specifically, the actual rebate amount is based on expected useful life of the measure(s) and annual kilowatt-hour (kWh) energy savings. Incentives range from $0.06 to $0.27 per kWh saved with the average incentive approximately $0.21/kWh.[16]

Smart Business Program Funding

Seattle City Light's Smart Business Program provides 100 percent funding for small businesses to replace inefficient lighting with approved energy-efficient lighting equipment. With interior lighting accounting for up to 60 percent of a typical business's energy bill, installation of efficient lighting can lower their energy and maintenance costs, improve lighting quality and increase productivity, and provide a safer working environment. Any small business in Seattle City Light's service area, not part of a chain, campus, or institution, is eligible for the programme.[17]

Upstream Non-Fiscal Tools

Seattle has implemented a variety of upstream non-fiscal tools to develop the circular economy and encourage a life cycle perspective to be taken by economic actors in an attempt to decouple economic growth from resource use and associated environmental impacts.

Green Building Permit Incentives: Priority Green Facilitated Incentive

Seattle's Department of Construction and Inspection offers the Priority Green Facilitated Incentive programme that streamlines the permitting process for master use permits in exchange for meeting green building standards. The city provides an opportunity to assist innovative projects that serve as visible models of high performance and sustainability. The programme provides developers with

- Reduced and predictable permit timelines that are established prior to the early design guidance application
- Priority processing for faster intake appointments, routing, and issuance
- Inter-departmental and integrated reviews to address citywide code challenges
- A single point of contact

To be eligible, projects require a signed agreement with the Department before the permit application is submitted. Furthermore, the project must achieve at least 10 points on the Priority Green Facilitated Building Matrix that lists which elements and points the project expects to achieve with a written description of how they will be achieved (Table 4.1). The completed Matrix must be embedded into the submitted plan set. As an alternative, applicants may submit a proposal that achieves a Leadership in Energy and Environmental Design (LEED) Platinum or Built Green five-star rating or meets the Living Building Challenge and complies with the 2030 Challenge.[18]

Green Building Permit Incentives: Priority Green Expedited

The Priority Green Expedited programme shortens the time it takes for developers to get a new construction permit in exchange for meeting a green building rating. The programme provides a single point of contact, prioritises the application, and provides faster initial review and routing of plans, and faster processing of the permit. To be eligible for the programme, the developer must be enrolled or registered with one of the following organisations and certification levels:

- Built Green® 4-Star, 5-Star, or Emerald Star
- LEED® Gold or Platinum

Table 4.1 Priority Green permitting

Aspect	Element	Description	Points
Energy and Climate Protection	Comply with 2030 Challenge (mandatory)	60 percent energy and fossil fuel use reduction using Energy Star Target Finder. For designs completed after 2015, the percentage moves up to 70 percent reduction.	*
	On-site Renewables	Minimum 2 percent of total energy use from any combination of wind, solar electric or solar thermal, biomass, etc.	1
	On-site and District Power Generation	Minimum 30 percent annual building energy load provided on-site, or minimum 30 percent annual energy building load supplied from district heating	1
	Passive Cooling Climate Responsive Design	Must provide at least two of the following: 1. Passive solar, thermal mass, etc. makes a minimum 10 percent contribution to total building energy use 2. Daylighting 75 percent of regularly occupied space 3. Natural or hybrid ventilation system serving more than 50 percent of regularly occupied space	1
	Other Innovative Energy and Climate Protection Elements	Provide proposal for alternative energy and climate strategies equivalent or greater in performance than, but not duplicating, the above	1–2

(continued)

Table 4.1 (continued)

Aspect	Element	Description	Points
Healthy People and Communities	Historic Landmark Development	Substantial retention of a building with historical characteristics	2
	Brownfield Development	Site is officially designated as brownfield or case can be made that it's in an area with environmental barriers to development	1
	Housing	50 percent or more of residential units are low-income housing. 75 percent of residential units are affordable to, and occupied by, households with incomes up to 120 percent of median income. 50 percent or more of units qualify as live/work	2
	Food Security	Food services including food banks, meal providers, or community kitchen. Produce food on sites physically covering area equivalent to 10 percent of site area	1
	Creates Green Collar Jobs	Detail type and number of green collar jobs equalling 50 percent or more of total jobs generated	1
	Innovative Transportation	Strategy to reduce single occupancy vehicles, enhance walkability, and create transit-friendly neighbourhoods	1
	Other Innovative Healthy People and Community Elements	Provide proposal for alternative strategies equivalent or greater in performance than, but not duplicating, the above	1–2
Restore our waters	Rainwater Use	50 percent or more of rainwater used on-site	1
	Grey Water/Black Water Reduction	50 percent or more reduction of grey/black water entering sanitary sewers	2
	Stormwater Infrastructure	Use of bioretention plants, swales, and stormwater dispersal resulting in a 75 percent reduction of stormwater flow from site	2
	Other Innovative Restore Our Waters Elements	Provide proposal for alternative strategies equivalent or greater in performance than, but not duplicating, the above	1–2

(continued)

Table 4.1 (continued)

Aspect	Element	Description	Points
Green Seattle Initiatives	Green Roof	50 percent, or more, green roof based on square footage of all roof surfaces	1
	Urban Forest	Planted trees provide canopy coverage of 25 percent of total site area	2
	Seattle Green Factor	Provide minimum coverage of 0.4 for commercial zones and 0.8 for multi-family zones	1
	Other Innovative Green Seattle Initiative Elements	Provide proposal for alternative strategies equivalent or greater in performance than, but not duplicating, the above	1–2
Waste Reduction and Recycling	Reuse, Recycling of Building Materials on Site	Retain minimum of 50 percent of existing buildings on site and/or reuse, recycle, or beneficially use 100 percent of asphalt, brick, and concrete and 50 percent of other building materials by weight. Reuse 20 percent of building materials by weight excluding asphalt, brick, and concrete	1
	Innovative Recycling	Recycling a minimum of 95 percent of construction, demolition, and land clearing waste by weight or volume	1
	Other Innovative Waste Reduction and Recycling Elements	Provide proposal for alternative strategies equivalent or greater in performance than, but not duplicating, the above	1–2

Total Green Building Matrix Summary Requirements

Comply with the 2030 Challenge
Achieve a minimum of 10 points
Include elements in three out of five environmental priority categories
Total Points:

Seattle Department of Planning and Development. 2016. Priority green permitting pilot program: Getting started. Available: https://www.seattle.gov/dpd/cs/groups/pan/@pan/documents/web_informational/dpds02376.pdf
*Compliance with 2030 Challenge must be discussed with Department of Planning and Development staff in an energy pre-submittal conference

- Living Building Challenge™, Petal or Net Zero Energy Building
- PASSIVE HOUSE INSTITUTE US +2015
- Seattle Department of Construction and Inspection Alternative Path (for small residential only)

Additionally, the developer must achieve Seattle's Priority Green requirements for energy and water conservation, waste reduction, stormwater management, and indoor air quality including

- At least 15 percent better than the 2012 Seattle Energy Code for non-residential and large-residential compared to the commercial provisions of the Seattle Energy Code or at least 20 percent better than the 2012 Seattle Energy Code for residential projects
- WaterSense plumbing fixtures or other comparable strategies to reduce water consumption[19]

Green Business Program

The Seattle Public Utilities' Green Business Program provides free tools and assistance to help Seattle businesses conserve resources and prevent pollution. Businesses are provided advice in the following:

- *Recycling, composting, and reducing waste*: The programme offers advice on how to ensure businesses comply with Seattle's new recycling and composting requirement laws, reduce carbon footprints, and improve their green image.
- *Preventing water pollution*: Businesses can access a variety of stormwater publications that can help prevent stormwater pollution and comply with stormwater regulations.
- *Saving water*: The programme offers advice on how businesses can reduce water use and lower utility costs by 20–30 percent by replacing inefficient equipment and changing operations.
- *Saving money*: The programme provides businesses with guidance on where to find rebates, useful publications, how-to guides, manuals, regulatory help, reports, and self-audit tools to green their business, comply with laws, and improve their bottom line.[20]

Green Business Program: Water Footprint Request

Seattle's Green Business Program enables non-residential users of water to request a water footprint analysis. Non-residential users can fill out a Water Footprint Request Form that includes information on their business location, water use, and any conservation opportunities already identified. The form is then reviewed by a Green Business Program representative who will then contact the customer to either provide assistance, collect more detailed data for a water footprint analysis, or arrange a site audit if necessary.[21]

Expert Energy Efficiency Advice for New Construction

Seattle City Light works collaboratively with design teams to create healthy, high-performance buildings with low, long-term operating costs. The utility's energy analysis assistance support helps designers build energy efficiency into the project from the beginning. Energy management analysts work with the design team to help optimise building systems for both efficiency and high performance.[22]

Seattle's Building Tune-up Requirement

Seattle requires owners of non-residential buildings 50,000 square feet or greater to tune-up their building's water and energy systems every five years. A tune-up includes

- An inspection of the building's system to identify operational and maintenance issues
- Corrections to operational issues identified in the inspection that have quick paybacks
- A report to the city Office of Sustainability and Environment summarising issues identified and actions taken

This requirement will be phased in for buildings that are incrementally larger than 50,000 square feet, summarised in Table 4.2. The dates indicate the deadline for the first required tune-up, from which buildings are required to repeat this schedule every five years.

Table 4.2 Seattle's building tune-up timeline

Building size	Due date of report on findings and actions taken
200,000 square feet or greater	1 October 2018
100,000–199,000 square feet or greater	1 October 2019
70,000–99,999 square feet or greater	1 October 2020
50,000–69,999 square feet or greater	1 October 2021

Seattle Office of Sustainability and Environment. Frequently asked questions: City of Seattle building tune-up requirement. Available: http://www.seattle.gov/Documents/Departments/OSE/Tune-Up_Website_FAQ.pdf

Downstream Fiscal Tools

Seattle has implemented a variety of downstream fiscal tools to develop the circular economy and encourage a life cycle perspective to be taken by economic actors in an attempt to decouple economic growth from resource use and associated environmental impacts.

Seattle's Commercial Compost Collection Service

Businesses that generate food waste or compostable paper must subscribe to a composting service or self-haul their waste to a transfer station for processing. Seattle businesses can save money and reduce waste through Seattle's Commercial Compost Collection Service which costs 32 percent less than regular garbage pickup prices, with food scraps and yard waste turned into compost.[23]

Downstream Non-Fiscal Tools

Seattle has implemented a variety of downstream non-fiscal tools to develop the circular economy and encourage a life cycle perspective to be taken by economic actors in an attempt to decouple economic growth from resource use and associated environmental impacts.

Ordinance Prohibiting Recyclables and Compostables in Garbage

Seattle prohibits residents and businesses from putting food scraps, compostable paper, yard waste, and recyclables in their garbage. Business owners and property managers must provide convenient food and yard waste services and recycling services at their property. Seattle Public Utilities gives warning notices for garbage containers that contain recyclables or compostables. For each warning, the property will receive a tag on the container and a notice will be mailed out to the account-holder. After two warnings, properties will receive a $50 fee on their waste bill for recyclables in the garbage.[24]

Green Business' Get on the Map

Seattle's Green Business' Internet-based Get on the Map campaign publicly recognises businesses that are taking actions to save water and energy and reduce waste and pollution. To be put on the map, businesses must take a minimum of five green actions from a list including having a spill kit on hand to prevent spills from entering stormwater drains; recycle and compost at the business; purchase and use recycled-content products; assign responsibility for environmental initiatives to a senior manager; measure and report the business's carbon footprint; buy products in bulk; install water-saving fixtures; and any other additional actions. Depending on the number of actions the business takes, the map icon will show up as green, greener, or greenest. The benefits of being on the map include

- *Gaining recognition*: The Green Business Program will promote the map online, through local media, and at community and business events.
- *Sharing success*: Businesses will receive materials touting their place on the map including website graphics, window clings, and a certificate from the city.

- *Receiving assistance*: Businesses can receive free one-on-one technical assistance to help them get on the map or darken their shade of green.[25]

Benchmarking Building Energy

In Seattle, buildings account for 33 percent of the city's emissions. The Benchmarking Building Energy policy supports Seattle's goal of reducing energy use and greenhouse gas emissions from existing buildings. As part of the policy, the city has an annual benchmarking, reporting, and disclosure regime for existing buildings. Building owners are required to benchmark their properties and authorise the City of Seattle to download annual energy performance data for each building. Since 2015, data for all buildings over 20,000 square feet are publicly available via a public-facing website. The city also requires that building owners or managers disclose a Statement of Energy Performance Report with tenants, buyers, or other qualified parties on request. Overall, the benefits of benchmarking include

- Showing property owners and managers how their buildings are using energy, which is the first step towards lowering energy costs and staying competitive
- Helping businesses and consumers make more informed decisions that take actual energy costs into account when buying or renting properties
- Lowering energy costs to owners and tenants, reducing greenhouse gas emissions, and creating jobs in the energy service sector
- Establishing energy performance ranges for Seattle building types based on their reported energy use to help building owners see how their building's energy use compares to their peers
- Allowing the city to track its energy reduction goals[26]

Business Waste Assessment Tool

Seattle offers a Business Waste Assessment Tool to help businesses reduce waste and increase recycling. The tool has a four-step process as follows.

1. *Plan*: Supplies and information needed to conduct a waste assessment

   a. Designate one or a few staff members to conduct the assessment

      i. Determine current dumpster size, pickup frequency, and monthly cost for garbage and recycling services
      ii. Gather all necessary supplies for conducting a waste assessment
      iii. Schedule a date and time to carry out a waste assessment, preferably right before the hauler pickup time

   b. Enter basic information including how many containers are used, number of pickups per week, monthly cost, and how full were the bins at pickup
   c. Obtain supplies for taking samples directly from the trash including rubber gloves, tape measure, camera, and blank fillable forms

2. *Gather data*: Observations about your current trash and recycling activities

   a. Print out the Business Waste Assessment form
   b. Estimate the percentages of each material (cardboard, office paper, plastic bags, food scraps, etc.) found in the trash, recycling, and compost bins

3. *Analyse results*: This section summarises the recycling potential and environmental benefits including reduced greenhouse gas emissions
4. *Act*: For instance, order free City of Seattle waste recycling posters for the business and set up a composting service[27]

Recycling Required for Construction and Demolition Projects

The City of Seattle has adopted a goal of recycling 70 percent of construction waste citywide by 2020. To achieve this, certain materials from new construction, remodelling, and demolition activities in

Seattle must be recycled and may not be put in containers for disposal in landfills. Before receiving a permit from Seattle's Department of Construction and Inspection, building permit applicants with projects that are more than 750 square feet in size and all demolition projects need to submit a Waste Diversion Report. In addition, a Salvage Assessment is to be filled out for whole building removal projects by a salvage verifier. Once the project is complete, all demolition permits and all new construction and remodelling projects that are $30,000 or more in value need to submit a Waste Diversion Report to Seattle Public Utilities. This report documents where project construction and demolition materials were delivered for reuse (on-site), salvage (off-site), recycling, and landfill disposal.[28]

Food Packaging Requirements

The City of Seattle requires all food service businesses to find recyclable or compostable packaging alternatives to all disposable food service items such as containers, cups, and other products. This applies to all food service businesses including restaurants, grocery stores, and institutional cafeterias. In addition, businesses with customer disposal stations where customers throw away single-use packaging must collect recyclable and compostable packaging in clearly labelled bins and sign up for composting and recycling services offered by their collection service provider.[29]

Case Study Summary

Seattle and the wider region's economy is led by its advanced aerospace and IT sector including cloud computing and software development. The State of Washington itself is one of the fastest-growing economies in the United States, and average household income in Seattle is well above the national average. Nonetheless, the city is experiencing a variety of challenges to its linear economy. Seattle has range of air quality issues including small particulates contributing significantly to ill-health in the population. Climate change will result in the region experiencing more frequent hot days and drier summers along with wetter fall, winter, and spring seasons. In addition, Seattle will experience more frequent storm events and storm surges. With Seattle dependent on hydropower, climate change variations in precipitation will reduce hydroelectric supply. As such, the city relies on a mix of fossil fuels and renewable energy for electricity generation. Seattle's population growth rate makes it one of the fastest-growing big cities in the country. The result of population growth will be an increase in waste generation over the course of the next decade. Despite a rising population, Seattle's total water demand is projected to decline in the long run; however, climate change will likely place stress on the water supply with changes in precipitation patterns resulting in competition between different water users.

Upstream

Seattle has implemented a variety of upstream fiscal and non-fiscal tools to develop the circular economy and encourage a life cycle perspective to be taken by economic actors in an attempt to decouple economic growth from resource use and associated environmental impacts. Some of the main tools are as follows:

- Seattle's water utility provides domestic and non-domestic customers with a variety of rebates to reduce water consumption, including rebates for toilets, commercial kitchen retrofits, and cooling tower improvements.
- Seattle's electricity utility provides rebates for small to medium-sized business customers to make energy efficiency upgrades with the rebate amount determined by the cost savings of the measures taken. In addition, the utility's Small Business Program provides funding for commercial customers to upgrade their lighting.

- Seattle's green building permit programme promotes the incorporation of green designs in buildings by giving developers who submit plans that meet LEED or other relevant certifications faster permitting times and priority processing of applications.
- Seattle provides free tools and assistance to help businesses conserve resources and prevent pollution with advice given in the areas of recycling, preventing water pollution, saving water, and how to save money from rebates and self-audits. In addition, businesses can request for a water footprint analysis to be conducted.
- Seattle's electricity provider works with designers to help them integrate energy efficiency into building projects from the beginning.
- The city requires all large non-residential buildings tune-up their water and energy systems every few years with the requirement phased in for buildings of certain sizes.

Downstream

Seattle has implemented a variety of downstream fiscal and non-fiscal tools to develop the circular economy and encourage a life cycle perspective to be taken by economic actors in an attempt to decouple economic growth from resource use and associated environmental impacts. Some of the main tools are as follows:

- Seattle mandates that all businesses that generate food waste or compostable paper subscribe to a composting service or have their waste taken to a transfer station for processing. To facilitate this requirement, the city offers a commercial compost collection service that costs less than regular services.
- Seattle has prohibited any residents or businesses from putting food scraps and recyclable materials in their rubbish with businesses and property managers required to provide convenient food and recyclable-related services at their properties.
- Seattle's online Get on the Map publicly recognises businesses that have acted to save water and energy and reduce waste and pollution.

To be put on the map, business must take a minimum amount of green actions. The more actions taken, the darker a business's icon appears on the map.

- Seattle requires all buildings of a certain size and over to benchmark their energy usage with the city making available data to the public. This enables businesses and consumers to take actual energy costs into account when buying or renting properties. Furthermore, the benchmark enables building owners to compare their energy performance with their peers.

- The Business Waste Assessment Tool helps businesses reduce their waste and increase recycling. The tool involves a four-step process in which businesses plan to conduct a waste audit, gather data, analyse the results, and act; for instance, order free City of Seattle waste recycling posters for the business.

- Seattle mandates that certain materials from new construction, remodelling, and demolition activities must be recycled. Before receiving a permit from the city's Department of Construction and Inspection, building permit applicants with projects over a certain size and all demolition projects must submit a Waste Diversion Report as well as a Salvage Assessment. Once the project is complete, projects over a specific monetary value will need to submit a Waste Diversion Report to Seattle Public Utilities.

- The City of Seattle requires all food service businesses to find recyclable or compostable packaging alternatives to all disposable food service items.

Overall, Seattle has implemented a variety of upstream and downstream fiscal and non-fiscal tools to develop the circular economy and encourage a life cycle perspective to be taken by economic actors in an attempt to decouple economic growth from resource use and associated environmental impacts. These tools are summarised in Table 4.3.

Table 4.3 Seattle case study summary

Tool	Tool type	Tool title	Description	Upstream/ Downstream
Fiscal	Environmental Taxes and Charges	Commercial Compost Collection Service	The city offers a commercial compost collection service that costs less than regular services	Downstream
	Subsidies and Incentives	Saving Water Partnership Rebates	Seattle's water utility provides domestic and non-domestic customers with a variety of rebates	Upstream
		Energy Efficiency Rebates for Industrial Customers	Rebates for business customers to make energy efficiency upgrades	Upstream
		Smart Business Program	Funding for commercial customers to upgrade their lighting	Upstream
Non-Fiscal	Regulations	Priority Green Facilitated Incentive	Developers who submit plans that meet LEED etc. receive faster permitting times and priority processing of applications	Upstream
		Priority Green Expedited	Shortens the time it takes for developers to get a new construction permit in exchange for meeting a green building rating	Upstream
		Building Tune-Up Requirement	The city requires all large non-residential buildings tune-up their water and energy systems every few years with the requirement phased in for buildings of certain sizes	Upstream
		Ordinance Prohibiting Recyclables and Compostables in Garbage	Seattle has prohibited any residents or businesses from putting food scraps and recyclable materials in their rubbish	Downstream
		Benchmarking Building Energy	All buildings of a certain size and over must benchmark their energy usage enabling businesses and consumers to make more informed decisions when buying or renting properties	Downstream
		Recycling Required for Construction and Demolition Projects	The city mandates that certain materials from new construction, remodelling, and demolition activities must be recycled	Downstream
		Food Packaging Requirements	All food service businesses must find recyclable or compostable packaging alternatives to all disposable food service items	Downstream
	Raising Industry Awareness and Capacity	Green Business Program	Free tools and assistance to help businesses conserve resources and prevent pollution	Upstream
		Water Footprint Request	Seattle businesses can request for the conducting of a water footprint analysis	Upstream
		Expert Energy Efficiency Advice for New Construction	Seattle's electricity provider works with designers to help them integrate energy efficiency into building projects from the beginning	Upstream
	Information-Based Tools	Get on the Map	An online map that publicly recognises businesses that have acted to save resources and reduce waste and pollution	Downstream
		Business Waste Assessment Tool	A four-step process helps businesses reduce their waste and increase recycling	Downstream

Notes

1. Trade Development Alliance of Greater Seattle. 2015. *Greater Seattle economy* [Online]. Available: http://www.seattletradealliance.com/economy/about-greater-seattle.
2. U.S. Bureau of Economic Analysis. 2017. Gross domestic product by state: Fourth quarter and annual 2016. Available: https://www.bea.gov/newsreleases/regional/gdp_state/2017/pdf/qgsp0517.pdf.
3. The Seattle Times. 2016a. *$80,000 median: Income gain in Seattle far outpaces other cities.* Available: https://www.seattletimes.com/seattle-news/data/80000-median-wage-income-gain-in-seattle-far-outpaces-othercities/.
4. The Seattle Times. 2016b. *Seattle area's air quality gets poor grades* [Online]. Available: http://www.seattletimes.com/seattle-news/seattle-areas-air-quality-gets-poor-grades/.
5. Washington State Department of Ecology. 2009. Health effects and economic impacts of fine particle pollution in Washington. Available: https://fortress.wa.gov/ecy/publications/documents/0902021.pdf.
6. Seattle Office of Sustainability and Environment. 2017a. Preparing for climate change. Available: https://www.seattle.gov/Documents/Departments/OSE/ClimateDocs/SEAClimatePreparedness_August2017.pdf.
7. City of Seattle. 2017a. *Fuel mix: How Seattle City Light electricity is generated* [Online]. Available: http://www.seattle.gov/light/FuelMix/.
8. U.S. EPA. 2017. *Climate impacts in the Northwest* [Online]. Available: https://www.epa.gov/climate-impacts/climate-impacts-northwest.
9. Office of Planning and Community Development. 2017. *About Seattle* [Online]. Available: http://www.seattle.gov/opcd/population-and-demographics/about-seattle.
10. The Seattle Times. 2017. *Seattle once again nation's fastest-growing big city; population exceeds 700,000* [Online]. Available: http://www.seattletimes.com/seattle-news/data/seattle-once-again-nations-fastest-growing-big-city-population-exceeds-700000/.
11. Seattle Public Utilities. 2013. Seattle solid waste trends. Available: http://www.seattle.gov/util/cs/groups/public/@spu/@garbage/documents/webcontent/02_015204.pdf.
12. City of Seattle. 2013. 2013 Water system plan: Plan summary. Available: http://www.seattle.gov/util/cs/groups/public/@spu/@water/documents/webcontent/04_007871.pdf.

13. City of Seattle. 2017b. *Long-range water demand forecast and firm yield estimate* [Online]. Available: http://www.seattle.gov/util/Documents/Plans/Water/DemandForecast/index.htm.

14. U.S. EPA. 2017. *Climate impacts in the Northwest* [Online]. Available: https://www.epa.gov/climate-impacts/climate-impacts-northwest.

15. Saving Water Partnership. 2017. *Commercial and industrial facilities* [Online]. Available: http://www.savingwater.org/Businesses/CommercialIndustrial/index.htm.

16. Seattle City Light. 2017a. *Funding your project* [Online]. Available: http://www.seattle.gov/light/business/incentives/default.asp.

17. Seattle City Light. 2017c. *Smart business program* [Online]. Available: http://www.seattle.gov/light/Conserve/business/cv5_sbiz.htm.

18. Seattle Department of Construction and Inspections. 2017b. Priority green facilitated. Available: http://www.seattle.gov/dpd/permits/greenbuildingincentives/prioritygreenfacilitated/default.htm.

19. Seattle Department of Construction and Inspections. 2017a. *Priority green expedited* [Online]. Available: http://www.seattle.gov/dpd/permits/greenbuildingincentives/prioritygreenexpedited/default.htm.

20. Seattle Public Utilities. 2017e. *Green your business* [Online]. Available: http://www.seattle.gov/Util/ForBusinesses/GreenYourBusiness/index.htm.

21. Seattle Public Utilities. 2017h. *Water footprint request form* [Online]. Available: http://www.seattle.gov/Util/ForBusinesses/GreenYourBusiness/SaveWater/WaterFootprintRequestForm/index.htm.

22. Seattle City Light. 2017b. *New construction* [Online]. Available: http://www.seattle.gov/light/conserve/business/cv5_ncp.htm.

23. Seattle Public Utilities. 2017c. *Commercial/Business customers* [Online]. Available: http://www.seattle.gov/Util/ForBusinesses/SolidWaste/FoodYardBusinesses/Commercial/index.htm.

24. Seattle Public Utilities. 2017b. *Clear alleys program rates* [Online]. Available: http://www.seattle.gov/Util/ForBusinesses/SolidWaste/GarbageBusinesses/Commercial/ClearAlleyProgram/Rates/index.htm.

25. Seattle Public Utilities. 2017a. *About the map* [Online]. Available: http://www.seattle.gov/util/ForBusinesses/GreenYourBusiness/GetontheMap/AbouttheMap/index.htm.

26. Seattle Office of Sustainability and Environment. 2017b. *Why benchmarking is required* [Online]. Available: https://www.seattle.gov/environment/buildings-and-energy/energy-benchmarking-and-reporting/why-benchmarking-is-required---about-the-law.

27. Seattle Public Utilities. 2017g. *Tools, resources and guides* [Online]. Available: http://www.seattle.gov/util/ForBusinesses/GreenYourBusiness/ToolsResourcesGuides/index.htm.

28. Seattle Public Utilities. 2017f. *Recycling required for construction and demolition project* [Online]. Available: http://www.seattle.gov/util/ForBusinesses/Construction/CDWasteManagement/RecyclingRequirements/index.htm.

29. Seattle Public Utilities. 2017d. *Food service packaging requirements* [Online]. Available: http://www.seattle.gov/util/ForBusinesses/Solid Waste/FoodYardBusinesses/Commercial/FoodPackagingRequirements/index.htm.

References

City of Seattle. 2013. 2013 Water system plan: Plan summary. Available: http://www.seattle.gov/util/cs/groups/public/@spu/@water/documents/webcontent/04_007871.pdf.

———. 2017a. *Fuel mix: How Seattle City Light electricity is generated* [Online]. Available: http://www.seattle.gov/light/FuelMix/.

———. 2017b. *Long-range water demand forecast and firm yield estimate* [Online]. Available: http://www.seattle.gov/util/Documents/Plans/Water/DemandForecast/index.htm.

Office of Planning and Community Development. 2017. *About Seattle* [Online]. Available: http://www.seattle.gov/opcd/population-and-demographics/about-seattle.

Saving Water Partnership. 2017. *Commercial and industrial facilities* [Online]. Available: http://www.savingwater.org/Businesses/CommercialIndustrial/index.htm.

Seattle City Light. 2017a. *Funding your project* [Online]. Available: http://www.seattle.gov/light/business/incentives/default.asp.

———. 2017b. *New construction* [Online]. Available: http://www.seattle.gov/light/conserve/business/cv5_ncp.htm.

———. 2017c. *Smart business program* [Online]. Available: http://www.seattle.gov/light/Conserve/business/cv5_sbiz.htm.

Seattle Department of Construction and Inspections. 2017a. *Priority green expedited* [Online]. Available: http://www.seattle.gov/dpd/permits/greenbuildingincentives/prioritygreenexpedited/default.htm.

———. 2017b. Priority green facilitated. Available: http://www.seattle.gov/dpd/permits/greenbuildingincentives/prioritygreenfacilitated/default.htm.

Seattle Department of Planning and Development. 2016. Priority green permmitting pilot program: Getting started. Available: https://www.seattle.gov/dpd/cs/groups/pan/@pan/documents/web_informational/dpds021376.pdf.

Seattle Office of Sustainability and Environment. 2017a. Preparing for climate change. Available: https://www.seattle.gov/Documents/Departments/OSE/ClimateDocs/SEAClimatePreparedness_August2017.pdf.

———. 2017b. *Why benchmarking is required* [Online]. Available: https://www.seattle.gov/environment/buildings-and-energy/energy-benchmarking-and-reporting/why-benchmarking-is-required---about-the-law.

———. Frequently asked questions: City of Seattle building tune-up requirement. Available: http://www.seattle.gov/Documents/Departments/OSE/Tune-Up_Website_FAQ.pdf.

Seattle Public Utilities. 2013. Seattle solid waste trends. Available: http://www.seattle.gov/util/cs/groups/public/@spu/@garbage/documents/webcontent/02_015204.pdf.

———. 2017a. *About the map* [Online]. Available: http://www.seattle.gov/util/ForBusinesses/GreenYourBusiness/GetontheMap/AbouttheMap/index.htm.

———. 2017b. *Clear alleys program rates* [Online]. Available: http://www.seattle.gov/Util/ForBusinesses/SolidWaste/GarbageBusinesses/Commercial/ClearAlleyProgram/Rates/index.htm.

———. 2017c. *Commercial/Business customers* [Online]. Available: http://www.seattle.gov/Util/ForBusinesses/SolidWaste/FoodYardBusinesses/Commercial/index.htm.

———. 2017d. *Food service packaging requirements* [Online]. Available: http://www.seattle.gov/util/ForBusinesses/SolidWaste/FoodYardBusinesses/Commercial/FoodPackagingRequirements/index.htm.

———. 2017e. *Green your business* [Online]. Available: http://www.seattle.gov/Util/ForBusinesses/GreenYourBusiness/index.htm.

———. 2017f. *Recycling required for construction and demolition project* [Online]. Available: http://www.seattle.gov/util/ForBusinesses/Construction/CDWasteManagement/RecyclingRequirements/index.htm.

———. 2017g. *Tools, resources and guides* [Online]. Available: http://www.seattle.gov/util/ForBusinesses/GreenYourBusiness/ToolsResourcesGuides/index.htm.

———. 2017h. *Water footprint request form* [Online]. Available: http://www.seattle.gov/Util/ForBusinesses/GreenYourBusiness/SaveWater/WaterFootprintRequestForm/index.htm.

The Seattle Times. 2016a. *$80,000 median: Income gain in Seattle far outpaces other cities*. Available: https://www.seattletimes.com/seattle-news/data/80000-median-wage-income-gain-in-seattle-far-outpaces-othercities/.

———. 2016b. *Seattle area's air quality gets poor grades* [Online]. Available: http://www.seattletimes.com/seattle-news/seattle-areas-air-quality-gets-poor-grades/.

———. 2017. *Seattle once again nation's fastest-growing big city; population exceeds 700,000* [Online]. Available: http://www.seattletimes.com/seattle-news/data/seattle-once-again-nations-fastest-growing-big-city-population-exceeds-700000/.

Trade Development Alliance of Greater Seattle. 2015. *Greater Seattle economy* [Online]. Available: http://www.seattletradealliance.com/economy/about-greater-seattle.

U.S. Bureau of Economic Analysis. 2017. Gross domestic product by state: Fourth quarter and annual 2016. Available: https://www.bea.gov/newsreleases/regional/gdp_state/2017/pdf/qgsp0517.pdf.

U.S. EPA. 2017. *Climate impacts in the Northwest* [Online]. Available: https://www.epa.gov/climate-impacts/climate-impacts-northwest.

Washington State Department of Ecology. 2009. Health effects and economic impacts of fine particle pollution in Washington. Available: https://fortress.wa.gov/ecy/publications/documents/0902021.pdf.

5

Natural Resource Management and the Circular Economy in Flanders

Introduction

Flanders covers nearly 45 percent of Belgium's territory and represents most of the country's industry and workforce with the region providing 58 percent of the national GDP. Flanders is an important logistic hub due to its central location and the region's dense integrated, multimodal transport infrastructure: over 800 European distribution centres are located there.[1,2] The region's GDP growth in 2016 was 1.4 percent, compared to the Brussels-Capital Region of 0.8 percent. In 2017, Flanders' growth rate is projected to be 1.7 percent and over the 2018–2021 period 1.6 percent.[3]

Challenges to the Linear Economy

Flanders is experiencing a variety of challenges to its traditional linear economy as described below through an assortment of examples.

© The Author(s) 2018
R. C. Brears, *Natural Resource Management and the Circular Economy*,
Palgrave Studies in Natural Resource Management,
https://doi.org/10.1007/978-3-319-71888-0_5

Air Pollution

In 2015, air quality results showed that Flanders met the European objectives for most pollutants; however, for some pollutants targets were not met. The concentrations of heavy metals in the air surrounding a couple of industrial plants exceeded European limits. When comparing air quality results from 2015 with the World Health Organization (WHO) guideline values for particulate matter, ozone, and sulphur dioxide, it was found that particulate matter was too high at nearly all measuring stations, ozone was too high at all measuring stations, and sulphur dioxide levels were too high at half of the measuring stations. There were also many sites where concentrations of nitrogen oxide and heavy metals were above WHO values.[4]

Climate Change

Flanders will likely experience an average temperature rise of between 0.7°C to 7.2°C over a period of 100 years. It is projected that the region will have an increase in precipitation in the winter months, which could be up to +38 percent over 100 years. This appears to be attributable to an increase in the amount of precipitation per day, rather than an increase in the number of wet days. Meanwhile, there will likely be a decrease in summer precipitation, with the decrease potentially amounting to −52 percent over 100 years due to a decrease in the number of wet days. Furthermore, it is predicted that during the summer months there will be periods of intense precipitation.[5]

Energy

Between 2009 and 2013, Flanders achieved a clear decoupling between economic growth and energy consumption. As a result, energy intensity of the Flemish economy decreased by almost 7 percent between 2000 and 2009. This change in intensity was due to structural effects (shifts in the importance of sectors within the Flemish economy) and changes in energy efficiency. This trend slowed during the global financial crisis with companies limiting new investments in energy-saving technology due to tighter constraints on obtaining credit. Today, energy intensity is 23 percent below the 2000 level; however, gross domestic energy consumption over the period 2000 to 2014 has decreased by only 7 percent. Regarding meeting energy needs of end users, the share of renewable energy with respect to energy consumption has increased from 5.7 percent in 2014 to 6 percent in 2015.[6] Concerning renewable energy, biomass has increased its share from 1 percent in 1990 to 4 percent in 2014.[7]

Population Growth

Each year, the population of Flanders grows by around 100,000 people. By 2030, Flanders will have a population of 7 million, compared to 6.4 million in 2013. Between 2005 and 2030, the population will have increased by 12 percent.[8]

Waste

The annual amount of collected residual waste in Flanders has decreased from 191 kilograms per person to 150 kilograms per person over the period 2000–2010. In 2015, residual waste per capita had decreased to 141 kilograms per person, which is a 57.4 percent decrease since 1991. Industrial waste has decreased by 19.2 percent since 2004. In addition, 77 percent of industrial waste is reused.[9]

Water

Water plays a vital role in the Flemish economy with water used in agricultural production, the chemical industry, power plants, refineries, steel industries, as well as for other high-value industries that require high-quality water. Sixty percent of total water consumption is used for cooling purposes. Excluding cooling water, households, industry, agriculture, and the energy sector consume 40 percent, 38 percent, 10 percent, and 6 percent of water respectively. Nonetheless, both the quality and availability of freshwater is under increasing pressure, with availability of water in Flanders and Brussels only 1700 cubic metres per capita. In fact, Flanders has less water availability than Spain, Portugal, and Greece.[10]

Upstream Fiscal Tools

Flanders has implemented a variety of upstream fiscal tools to develop the circular economy and encourage a life cycle perspective to be taken by economic actors in an attempt to decouple economic growth from resource use and associated environmental impacts.

Ecological Investment Grants

Flanders Innovation and Entrepreneurship (VLAIO) has a range of investment grants to support ecological investments as follows.

Ecology Premium-Plus Investment Subsidy

The Ecology Premium-Plus (EP-Plus) investment subsidy is for all enterprises established in Flanders. The subsidy is calculated based on the supplementary investment cost of the eligible investment components up to EUR 1 million over a period of three years, with grants only made to technologies on a limited technology list that comprises environmental technologies, energy technologies, and renewable energy. The grant size depends on the size of the company, the performance of the technology, and the type of technology used, with up to 25 percent granted for environmental investments (12.5 percent for large enterprises). Subsidies can be increased with a subsidy bonus of up to 10 percent for enterprises that hold environmental certificates or an environmental management system.

Strategic Ecology Support Grant

The Strategic Ecology Support (STRES) grant is for companies using technologies which are, due to their exception and unique character, not on the limited technology list. The investment grant is intended for strategic environmental projects that contribute to global solutions (environmental or energy issues); focus on closed circuits (renewable energy, sustainable use of material, and recuperation of material); and feature processes involving integrated solutions. Companies must also have invested a minimum of EUR 3 million. The size of the grant depends on the type of investment, the performance of the technology, and the size of the company, with support for small and medium-sized enterprises (SMEs) amounting to as much as 40 percent (30 percent for large enterprises). Overall the grant amount is limited to a maximum of EUR 1 million every three years. Nonetheless, before a grant can be approved, a feasibility study must be undertaken.[11]

R&D Investment Tax Breaks

Flanders offers tax deductions and credits to foreign research and development (R&D)-intensive companies as follows.

Investment Deduction

The Investment Deduction tax break is for fixed assets (non-patents) that aim to promote R&D of new products and advanced technologies that are environmentally friendly. The deduction is 13.5 percent on the investment (at once) or 20.5 percent of the annual depreciation (staggered deduction). For patents acquired or self-developed by the company, the deduction is 13.5 percent on the investment value (staggered deduction is not offered).[12]

2bis Tax Credit

The 2bis Tax Credit enables companies that have not made an investment deduction over five consecutive years to have a guaranteed cash reimbursement that can be carried forward. This credit allows a company to have the same benefit as an investment deduction even in the case of an insufficient taxable basis.[13]

R&D Investment Deduction: High-Tech Innovation

To stimulate the local economy and encourage the production of high-tech products, Flanders awards an R&D investment deduction to companies investing in new high-tech assets. The deduction, available to companies in any sector that invest in new assets used to produce high-tech products (there is no formal definition established for assets used), is 17 percent at once, or 20.5 percent of the annual investment (staggered deduction).[14]

Upstream Non-Fiscal Tools

Flanders has implemented a variety of upstream non-fiscal tools to develop the circular economy and encourage a life cycle perspective to be taken by economic actors in an attempt to decouple economic growth from resource use and associated environmental impacts.

Environmental Permit

Any company establishing itself in Flanders must determine on its own accord whether an environmental permit is required. This depends on whether the company falls into Category 1, 2, or 3 of the Flemish Environmental Permitting Regulations. For a Category 1 or 2 company, an environmental permit is required, while a Category 3 company must only report their activities. The companies themselves must determine which category they fall in: Categories 1, 2, and 3 refer to the level of nuisance the company can possibly cause to humans and the environment. Factors including size of storage, machinery used, and use of hazardous products or raw materials determine a company's classification:

- Category 1 refers to companies having the highest nuisance level.
- Category 2 refers to companies causing less nuisance.
- Category 3 refers to companies with the lowest level of nuisance.[15]

Ecolizer

The Public Waste Agency of Flanders (OVAM) has created the Ecolizer which is an eco-design tool that is aimed at designers and companies who wish to know and address the environmental impacts of their products. The Ecolizer can quickly and easily calculate the environmental impact of products, where impact is calculated based on all materials, processes, packaging, and transport modes that are used in the life cycle of a product. In addition to calculating the total environmental impact, Ecolizer can calculate the impacts of each phase in the life cycle of a product. This enables the targeting of specific life cycle phases that have high environmental impact. Furthermore, the tool enables users to compare their scores with other products.[16]

OVAM SIS Toolkit

The OVAM Sustainable Innovation System (SIS) Toolkit provides guidance to users on how to integrate sustainability principles with innovation and design processes and create sustainable value. The Toolkit includes a set of cards, three brainstorming posters (combined into a matrix in Table 5.1), and two dice that can be used with colleagues in a brainstorming session or design process. The matrix enables users to visualise new opportunities through a broad view of sustainability, with the matrix putting five forms of value creation (human, intellectual, financial, social, and natural capital) against a strategic/function perspective (ambition and needs) and life cycle perspective. The result is a matrix containing 30 leading questions spanning the whole spectrum of sustainable design. The definitions of each value used are as follows:

- *Human capital*: This is the totality of competencies and personal characteristics that allow individuals to develop themselves, realise personal goals, and engage in meaningful (professional) activities
- *Intellectual capital*: This is an individual's or organisation's stock of knowledge that yields a stream of useful products and/or services over time

Table 5.1 The OVAM Sustainable Innovation System matrix

	1.0 Human capital	2.0 Intellectual capital	3.0 Financial capital	4.0 Social capital	5.0 Natural capital
Ambition	1.1 How fundamentally do I (as a designer or client) want to contribute to the well-being and quality of life of the envisaged product/service's end user?	2.1 What role in our organisation do we, as a company, want to associate with intellectual capital and innovation?	3.1 What is the balance between business risk and opportunity that we, as a company, want to strike with the development of new products and services?	4.1 What is our ambition—as designer or client company—about the contribution of our products/ services to the social capital of users and their communities?	5.1 What is our ambition—as designers or clients—about the contribution of our products/ services to the protection and enhancement of natural capital?
Needs	1.2 Which human needs do I (as a designer/client/ company) want to fulfil with the envisaged product/service?	2.2 How can we, as a client company or design team, feed our intellectual capital through a better understanding of the current and future needs that our products are intended to serve?	3.2 How can we reinforce our business success through a better understanding of the needs which our products and services are designed to serve?	4.2 What kind of social capital-oriented needs can we address with our products and services?	5.2 What sustainability-oriented needs do we want to cater for with our envisaged products and services?

(continued)

Table 5.1 (continued)

	1.0 Human capital	2.0 Intellectual capital	3.0 Financial capital	4.0 Social capital	5.0 Natural capital
Design Process	1.3 How can a design process maximally contribute to the potential, competencies, and opportunities of the user?	2.3 How can a design process be shaped so that it maximally contributes to the organisation's intellectual capital?	3.3 How do we want to shape our business model to sustainably pursue financial success?	4.3 How can a design process contribute to users' and communities' social capital?	5.3 How can we, as designers, integrate the principles of eco-design into our work?
Production	1.4 How can a product/service be designed so that, in the production phase, co-workers' or suppliers' human capital is maximised?	2.4 How can a production process be shaped so that it maximally contributes to the intellectual capital of an organisation?	3.4 How can we produce our goods and services to maximally contribute to our financial capital?	4.4 How can a production process be shaped that maximally contributes to the social capital of those involved in it?	5.4 How can a production process be designed so that it has minimal negative impact on natural capital?

(continued)

Table 5.1 (continued)

	1.0 Human capital	2.0 Intellectual capital	3.0 Financial capital	4.0 Social capital	5.0 Natural capital
Distribution and Use	1.5 How can a product/service be designed so that, during distribution and use phases, users' human capital is maximised?	2.5 How can a distribution process be shaped so that it maximally contributes to the intellectual capital of a user or organisation?	3.5 How can we design products and services (and their distribution) so that their use contributes to our client's financial capital?	4.5 How can a product or service be designed so that its distribution and use contributes to community building and trust?	5.5 How can products and services be designed so that their distribution and use have minimum negative impact on the environment?
End of Life	1.6 How can a product/service be designed so that, at the end of its life, it still contributes to human capital?	2.6 How can a product be designed so that, at the end of its life cycle, it contributes to the user's or producer's intellectual capital?	3.6 How can we design our goods and services so that they even contribute to financial capital at the end of their life?	4.6 How can products be designed to contribute to social capital at the end of their life cycle?	5.6 How can we design products and services that at the end of their life cycle do not place an extra burden on natural capital?

OVAM. 2011. SIS brainstorm poster. Available: http://www.ecodesignlink.com/images/filelib/OVAMSISBrainstormposterEN2011m10d12_5267.pdf

- *Financial capital*: This reflects a company's or individual's economic resources and material well-being. It is purchasing power in the form of money available for the production or purchasing of goods and services
- *Social capital*: This includes any value added to the activities and economic outputs of an organisation by human relationships, partnerships, and cooperation
- *Natural capital*: This includes the natural resources (energy and matter) and processes needed by organisations to produce their products and deliver their services

Overall, the Toolkit:

- Offers creative support for people looking for innovative and sustainable solutions in all aspects of entrepreneurship and design
- Provides a clear, wide-angle perspective on sustainability seen through the lens of a clearly structured and user-friendly conceptual model
- Helps to better integrate different aspects of sustainability in an early stage of the design process
- Helps companies communicate with all stakeholders
- Contributes to any innovation or design process, whether it concerns the development of products, services, product-service concepts, environments or architecture[17,18]

Eco-clusters: Promoting Industrial Symbiosis

OVAM is building eco-clusters that focus on materials recovery. This is part of Flanders' goal of achieving 15 percent less waste from companies. The clusters encourage cooperation between companies within the same sector or across sectors, so materials can be kept in the loop. There are three specific types of eco-clusters OVAM promotes:

1. Collective (selective) collection of waste streams
2. Efficient use of raw materials and the application of eco-design strategies
3. Exchange of residues/industrial symbiosis[19]

The Symbiosis Project

The Symbiosis Project is a matchmaking platform designed to provide a high-quality application of industrial waste and waste streams. With funding from VLAIO and OVAM, the Symbiosis Project team facilitates matchmaking through active workshops, company visits, and an interactive database. In addition, the programme assists in removing potential barriers.[20]

Rewarding Mini-Enterprises

An important group of aspiring entrepreneurs in the circular economy is pupils and students. OVAM is cooperating with Vlajo, an association that aims to familiarise young students with entrepreneurship and integrate eco-efficiency in the enterprises of tomorrow. Each year both organisations select one mini-company that excels in eco-efficiency and reward it with an Eco-Efficiency Award.[21]

Material Scan

OVAM, with financial support from VLAIO, provided a material scan service for over 120 SMEs with the results reviewed by specialist consultants. The first scan was intended to sensitise the businesses and encourage them to look at ways of reducing raw material usage. The scan's results revealed that raw material losses could be reduced by an average of no less than 10 percent and that 2 percent could be saved from production costs.[22]

Eco-Efficiency Scan

The Eco-Efficiency Scan is an Internet application that enables organisations to detect eco-efficiency opportunities within the company. It is a simple, quick tool that provides users with insights on possible measures that can increase the eco-efficiency of any business (both production and service). The Eco-Efficiency Scan, which is completely anonymous, consists of:

- A brief analysis of the company.
- An overview of eco-efficiency measures. These measures are grouped by topics with users able to choose which measures are most relevant, with the choice recorded in the report.
- The possibility of calculating many indicators.
- A reporting module that examines the options selected. Here, the user can receive advice on how to get started or seek referrals for more specialised advice.[23]

Downstream Fiscal Tools

Flanders has implemented a variety of downstream fiscal tools to develop the circular economy and encourage a life cycle perspective to be taken by economic actors in an attempt to decouple economic growth from resource use and associated environmental impacts.

Green Event Scan

The Green Event Scan is an online tool that helps event organisers screen their events on sustainability. With simple clicks, users can scan through all aspects of their events—waste, energy, catering, space, mobility, and materials use—and receive scores along with tips for improvements in each of the areas. Users also receive a visual representation of their score with a scan rose that can be easily shared on social media.

The GreenVent Award

Users of the Green Event Scan can win the GreenVent Award along with a cash prize. The prize pool of EUR 10,000 is split among the winners of three categories:

- *Small events (500 to 3000 visitors)*: EUR 1000
- *Medium events (3000 to 15,000 visitors)*: EUR 3000
- *Major events (over 15,000 visitors)*: EUR 6000[24]

A jury of experts from the environmental and events sector will assess each applicant's dossier. The jury's decision is based on the scores from the Green Event Scan and considers the efforts of the organisers and their visitors.[25]

Downstream Non-Fiscal Tools

Flanders has implemented a variety of downstream non-fiscal tools to develop the circular economy and encourage a life cycle perspective to be taken by economic actors in an attempt to decouple economic growth from resource use and associated environmental impacts.

The Eco-design in Higher Education Kit

The Eco-design in Higher Education (EHE) Kit provides guidance to teachers, professors, education coordinators, and training councils on integrating the concept of eco-design in higher education programmes. The EHE Kit consists of an eco-design in higher education guide and three specific and concise information cards that have a workable format.

1. *Learning content cards*: These cards describe the various relevant themes connected with eco-design, for example, biomimicry, eco-labels, life cycle thinking, and so on. Each card provides a short description of the theme, the most relevant references on the theme, and a link to the practical example cards.
2. *Teaching method cards*: These cards describe teaching methods that can be applied for integrating eco-design in a higher education programme, for example, brainstorming, demonstration, and group work. Each card provides a short description of the teaching method

and provides references to a more extensive explanation of the teaching method and a link to practical examples (practical example cards).

3. *Practical example cards*: These cards provide practical examples for specific learning content and teaching methods and can serve as inspiration for applying eco-design in one's own programme, for example, a debate on depletion of raw materials or becoming aware of the life cycle of products. Each card describes the practical example, the learning content provided, development of competencies, the applicable teaching methods, and possible points of attention and refers to the programme to which the example is provided.[26,27,28]

Plan 2016–2020 Public

The Flemish Government aims for 100 percent sustainable procurement by 2020. In support, the Plan 2016–2020 Public was created with an emphasis on professionalism, innovation, sustainability, and access to SMEs. The Plan focuses on six themes:

1. *Professionalising procurement*: A professional procurement contributes to achieving the goals of the Flemish Government. In this context, professional procurement is based on investments in the areas of knowledge about buying work, supplies and services, market conditions, and applicable regulations. Attention is also paid to the life cycle costs of products and services procured.

2. *Achieving end-to-end e-procurement*: Tenders for public contracts are to be submitted digitally via the online platform e-tendering. The overall purchase process is easily subjected to systematic quality measurement and improvement, ensuring public costs are kept down.

3. *Promoting SME participation*: The Plan contains a range of actions and measures to ensure the participation of SMEs. For instance, the government will examine whether the solutions provided by SMEs for contracts offer a high score in terms of innovation and sustainability. This ensures that any diseconomies of scale are offset by SMEs' progressiveness in terms of sustainability and innovation.

4. *Integrity of public procurement*: The government will focus on ethical procurement, both in terms of the contracting authorities and of the tenderers or candidates. This is to ensure that the risk of waste, as well as fraud and corruption, is minimised.
5. *Sustainability and innovation*: The contracting authority will integrate environmental, social, and economic criteria in all phases of its contracts. With innovative procurement, it allows the contracting authority to develop new innovations or smart requirements in the specifications to encourage solutions that have the least impact on the environment throughout their life cycle.
6. *Support for strategic projects*: In the context of public-private partnerships, the government works with an output-orientated approach with the government formulating specific objectives, highlighting constraints, and the private sector working out practical solutions to meet the objectives. This enables participants to come up with their own sustainability solutions and innovative approaches.[29,30,31,32,33,34,35]

Material-Conscious Build Cycles

The Material-Conscious Build Cycles programme involves a permanent framework for cooperation between the government and construction industry to shape the sustainable management of materials from the recycling area, with efforts placed around five main themes:

1. *Management and regulation of recycled granulates*: The aim of this theme is to ensure the quality and traceability of recycled aggregates. This will be achieved by analysing how waste is generated, how it is collected, how it is transported, and how the acceptance of the debris in the crushing plants is done. Attention is also placed on the processing of rubble, the transport of recycled granules, and its effective use.
2. *Build dynamic or change-oriented buildings*: Buildings in the future will need to be able to respond to a socially, economically, and physically changed environment, for instance, meeting higher levels of energy efficiency. If buildings built today cannot respond to these

changes then it will lead to high financial costs, environmental nuisance, and new living or working environments until the buildings are ready for use, resulting in more transport, resource consumption, and waste.

3. *Selective demolition and dismantling*: Most of the waste in the coming years will be from stony rubble that contains materials such as spray insulation and composite materials. The quality of the materials collected determines whether they can be used subsequently for the construction of new buildings or in other applications. By selectively demolishing and dismantling, construction waste can be separated on site enabling better waste management and recycling.

4. *Non-stony cycle*: When demolishing buildings and infrastructure there is also non-stony materials including mineral insulation, plastic, plaster and plasterboard, roof bitumen, and metal left over. To ensure these materials can be recycled, logistics and cooperation across the material chain needs to be optimised.

5. *Stony cycle*: Flanders has a high material intensity (the use of materials converted per capita is among the highest in the European Union), but it is a resource-poor region and therefore sensitive to supply shortages and price fluctuations in the commodities market. As such, pure debris from the demolition or renovation of buildings or the breaking up of roads can be processed into recycled aggregates, thus closing the material cycles.[36,37,38,39,40]

Green Economy Covenant

The Green Economy Covenant (GEC) is a voluntary agreement between the private sector and the Flemish Government to start green projects together. As part of the GEC, environmental goals are to be pursued together with increased competition and good business practice. The GEC does not set targets unless specified by the parties. As part of a GEC, parties commit themselves to realising a green project within a three-to-four-year timeframe. The experience and knowledge gained from the green project is then disseminated in a final report which is published on the Green Deal website.[41]

Green Deal Circular Purchases

The Green Deal Circular Purchases initiative was formed in 2017 between the Flemish Government and The Shift, VVSG, and Bond Better Environment. Over a two-year period, the initiative will help create circular economy acquisition projects, develop knowledge about circular purchasing, and share and disseminate this knowledge. The initiative is open for others to join including purchasers who wish to start a circular economy acquisition project in their organisation and facilitators who wish to contribute to supporting the initiative.[42] As part of the initiative, purchasers need to meet the following requirements:

- Commit themselves to launching two circular economy purchases during the first year of the initiative (between June 2017 and June 2018)
- At the end of the two-year project (June 2019), purchasers need to show that they have considered the principles of the circular economy in their purchasing decisions and monitored the performance of their supplier(s)
- Participate in the meetings of the learning network set up by facilitating parties
- Knowledge and experience gained in the field of circular purchases needs to be disseminated via website, newsletter, and so on, and shared with the learning network

Throughout the process, participants are guided by group facilitators who organise a series of events, as well as provide tools and expertise. For instance, the facilitators:

- Aggregate knowledge and experience and share it throughout the network
- Organise workshops and meetings to allow participants to exchange experiences and help them launch and implement circular economy purchases

- Advise participants on methods and tools that can be used to incorporate circularity in procurement processes
- Provide lessons and experiences from the various pilot projects for participants to learn from
- Work with government agencies and business associations to try and remove bottlenecks and barriers to circular inventory purchases[43]

Master Circular Economy

The Master Circular Economy comprises four half-day sessions in which participants will gain an understanding of the principles, policies, and drivers of the circular economy. Some of the key questions the sessions (summarised in Table 5.2) will answer include what circular economy business strategies are available? And how can the circular economy offer various ecological and business opportunities?[44]

Case Study Summary

Flanders covers less than half of Belgium's territory yet contributes over 50 percent of the national GDP. The region's economic growth is higher than the national rate and is projected to continue growing over

Table 5.2 Master Circular Economy sessions

Day	Theme	Topics covered
1	Introduction: Understanding the Circular Economy	• Introduction • Strategy and business models
2	Circular Economy and Government	• Europe • Flanders • Local government
3	Key Drivers within a Successful Circular Economy	• Financing • Open innovation
4	Key Drivers within a Successful Circular Economy	• The role of smart technology • Logistics

Flanders Circular. 2017b. *Masterclass circular economics* [Online]. Available: http://www.vlaamsmaterialenprogramma.be/fmp

the next several years. Nonetheless, Flanders is experiencing a variety of challenges to its linear economy. While air pollution statistics show the region is meeting European objectives for most pollutants, some pollutant targets have not been met, with concentrations of heavy metals in the air surrounding some industrial plants exceeding European limits. In addition, the region has particulate matter levels that exceed WHO guideline values as do heavy metals. Climate change will see heavy precipitation levels during the winter months and low levels during the warmer months; however, during the summer months there will be periods of intense rainfall. While energy intensity is significantly below 2000 levels, there has only been a modest decrease in gross domestic energy consumption since the 2000s while renewable energy levels have only slightly increased over the past couple of years. The region's population is projected to be 12 percent higher in 2030 compared to 2005. Water quality and availability is coming under increasing pressure, with the region having less water availability than dry, arid regions of Europe.

Upstream

Flanders has implemented a variety of upstream fiscal and non-fiscal tools to develop the circular economy and encourage a life cycle perspective to be taken by economic actors in an attempt to decouple economic growth from resource use and associated environmental impacts. Some of the main tools are as follows:

- Flanders has a range of investment grants that support ecological investments. The EP-Plus subsidy is available to all enterprises in the region to implement environmental technologies as well as energy and renewable energy technologies that are on a limited technology list, with the size of the subsidy being dependent on the technology's performance.
- The STRES grant is available for companies that are using technology not on the limited technology list for strategic environmental projects that provide global solutions to environmental or energy challenges,

focus on closing the loop, and feature integrated solutions. The size of the grant is dependent on the performance of the technology.

- Flanders offers investment deduction tax breaks for companies to develop new environment-friendly products and technologies. Meanwhile, the High-Tech Innovation tax break is for companies in any sector that wish to invest in new assets to produce high-tech products.
- Any company that sets itself up in Flanders must check to see whether they require an environmental permit. Whether a permit is required is based on which category the company falls into, with companies having high levels of nuisance to humans and the environment requiring a permit.
- Flanders has developed the Ecolizer eco-design tool to help designers and companies calculate quickly and easily the environmental impact a product has throughout its life cycle.
- The OVAM SIS Toolkit has been developed to help users visualise new sustainability opportunities that can be integrated into the innovation and design process of products and services
- Flanders is developing a range of eco-clusters that will contribute to the region's aim of decreasing waste generated from companies by utilising waste streams, using raw materials more efficiently, and exchanging residues between companies.
- Flanders has developed the Symbiosis Project that matches companies' waste streams with other companies seeking to turn waste into resources. The project involves active workshops, site visits, and an interactive database.
- To enhance awareness of the circular economy among students, OVAM and Vlajo each year award an Eco-Efficiency Award to the best young entrepreneur in the region.
- OVAM with support from VLAIO provided a material scan for SMEs to encourage them to look at ways of reducing raw material usage with the results reviewed by specialist consultants. The scan's overall results provided evidence to SMEs that sustainable materials management can simultaneously provide significant economic and environmental benefits.

- The Eco-Efficiency Scan is an Internet application that enables any type of organisation to quickly identify measures that can increase eco-efficiency. The scan also directs users to where they can get specialised advice to increase their efficiency.

Downstream

Flanders has implemented a variety of downstream fiscal and non-fiscal tools to develop the circular economy and encourage a life cycle perspective to be taken by economic actors in an attempt to decouple economic growth from resource use and associated environmental impacts. Some of the main tools are as follows:

- The Green Event Scan tool helps event organisers reduce material and resource usage with users receiving tips on how to make improvements. Users also receive a visual representation of their score that can be shared on social media.
- Users of the Green Event Scan can win the GreenVent Award that includes a cash prize that is split among winners of the three categories of small, medium, and major-sized events.
- The EHE Kit provides guidance to teachers and other educators on how to integrate eco-design in higher education programmes. The kit includes a guidebook as well as informational cards on various aspects of the circular economy, teaching method cards to help instructors integrate eco-design thinking into courses, and practical example cards.
- In support of the Flemish Government's aim of 100 percent sustainable procurement by 2020, a plan has been developed that focuses on ensuring that sustainability aims are incorporated into the process, with a focus on life cycle thinking, as well as supporting public-private partnerships to encourage sustainability solutions and innovative circular economy approaches.
- The Flemish Government has developed a framework for cooperation between the government and construction industry to promote the

sustainable management of materials with a focus on managing and regulating recycled granulates, promoting buildings that respond to changing social, economic, and physical environments in the future, and ensuring waste is of a high quality for recycling.

- The Flemish Government's GEC is a voluntary agreement between the private sector and the government to start green projects together. The GEC involves all parties committing themselves to doing everything within their means to realise a project within a specific timeframe.
- The Green Deal Circular Purchases initiative commits participants to increase the level of circular purchases being made. Over a two-year period, participants will create circular acquisition projects, develop knowledge about circular purchasing, and share and disseminate this knowledge with other participants.
- Master Circular Economy is a series of half-day courses for participants to attend in which they will come away with knowledge on what circular economy business strategies are available and how the circular economy can provide numerous ecological and business opportunities.

Overall, Flanders has implemented a variety of upstream and downstream fiscal and non-fiscal tools to develop the circular economy and encourage a life cycle perspective to be taken by economic actors in an attempt to decouple economic growth from resource use and associated environmental impacts. These tools are summarised in Table 5.3.

Table 5.3 Flanders case study summary

Tool	Tool type	Tool title	Description	Upstream/ Downstream
Fiscal	Environmental Taxes and Charges	Investment Deduction	Tax breaks for companies to develop new circular economy products and technologies	Upstream
		2bis Tax Credit	Tax breaks to be carried forward by companies to develop new circular economy products and technologies	Upstream
		R&D Investment Deduction: High-Tech Innovation	Available for companies in any sector wishing to invest in new assets that are used to produce high-tech products	Upstream
	Subsidies and Incentives	EP-Plus	Available to all enterprises in the region to implement a range of environmental, energy, and renewable energy technologies	Upstream
		STRES Grant	Available for companies that are using technology not on the limited technology list for circular economy projects that address global challenges	Upstream
Non-Fiscal	Regulations	Environmental Permit	Any company that sets itself up in Flanders must check to see whether they require an environmental permit	Upstream
	Green Public Procurement	Plan 2016–2020 Public	The Plan focuses on life cycle thinking and innovative circular economy approaches to public procurement	Downstream
	Cluster Policies	Eco-clusters	Eco-clusters are being developed to build the region's circular economy	Upstream
		Symbiosis Project	Matches companies' waste streams with other companies seeking to turn waste into resources	Upstream

(continued)

Table 5.3 (continued)

Tool	Tool type	Tool title	Description	Upstream/ Downstream
	Education and Training	EHE Kit	Guides teachers and other educators on how to integrate eco-design in higher education programmes	Downstream
		Master Circular Economy	Course participants learn about circular economy business strategies	Downstream
	Raising Industry Awareness and Capacity	Material Scan	Encouraged SMEs to look at ways of reducing raw material usage	Upstream
		Eco-Efficiency Scan	A web application that enables organisations to detect eco-efficiency opportunities within the company	Upstream
		Green Event Scan	Helps event organisers screen their event on sustainability	Downstream
	Industry-Based Standards	GEC	A voluntary agreement between the private sector and the government to start green projects together	Downstream
	Voluntary Agreements	Green Deal Circular Purchases	Initiative commits participants to increase the level of circular purchases being made	Downstream
	Supporting Life-Cycle Analysis	Ecolizer	Helps calculate quickly and easily the environmental impact a product has throughout its life cycle	Upstream
		OVAM SIS Toolkit	Helps users visualise new sustainability opportunities that can be integrated into the innovation and design process of products and services	Upstream
	Environmental Recognition Awards	Rewarding Mini-Enterprises	Each year an Eco-Efficiency Award is given to the best young entrepreneur in the region	Upstream
		GreenVent Award	Winners of the Green Events Scan receive a GreenVent Award along with a cash prize	Downstream
	Knowledge Transfer Networks	Material-Conscious Build Cycles	The government and construction industry cooperate on promoting the sustainable management of materials including recycling	Downstream
	Information-Based Tools	The Eco-Efficiency Scan	Enables an organisation to quickly identify measures that can increase eco-efficiency	Upstream

Notes

1. European Commission. 2017. *Flanders* [Online]. Available: https://ec.europa.eu/growth/tools-databases/regional-innovation-monitor/base-profile/flanders.
2. Flanders Investment and Trade. 2017d. *Why Flanders is a trade powerhouse* [Online]. Available: https://www.flandersinvestmentandtrade.com/invest/en/investing-in-flanders/why-flanders-trade-powerhouse.
3. The Brussels Times. 2016. *Regional economic prospects for 2016–2021 – Slightly lower growth expected in Wallonia than in Flanders* [Online]. Available: http://www.brusselstimes.com/belgium/5964/regional-economic-prospects-for-2016-2021-slightly-lower-growth-expected-in-wallonia-than-in-flanders.
4. Flanders Environment Agency. 2017a. Air quality in the Flemish region 2015. Available: https://en.vmm.be/publications/air-quality-in-the-flemish-region-2015.
5. Flanders Environment Agency. 2015. MIRA climate report 2015: About observed and future climate changes in Flanders and Belgium. Available: http://www.milieurapport.be/en/publications/topic-reports/mira-climate-report-2015-about-observed-and-future-climate-changes-in-flanders-and-belgium/.
6. EnergyVille. 2016. *Inventory renewables Flanders 2005–2015* [Online]. Available: http://www.energyville.be/en/nieuwsbericht/inventory-renewables-flanders-2005-2015.
7. Flanders Environment Agency. 2017b. MIRA system balance 2017: Environmental challenges for the energy, mobility and food systems in Flanders. Available: http://milieurapport.be/en/publications/other-reports/mira-system-balance-2017-environmental-challenges-for-the-energy-mobility-and-food-systems.
8. Flanders Environment Agency. 2014. Megatrends: far-reaching, but also out of reach? How do megatrends influence the environment in Flanders? MIRA Future Outlook Report 2014. Available: https://en.vmm.be/news/archive/megatrends-far-reaching-but-also-out-of-reach.
9. OVAM. 2016. Activities report 2016. Available: http://www.ovamenglish.be/sites/default/files/atoms/files/OVAM_activiteitenoverzicht_2016_DRUK_EN_LR_SPREAD.pdf.
10. Vlakwa. 2015. The socio-economic importance of water in Flanders. Available: http://www.vlakwa.be/fileadmin/media/pdf/20150605_samenvatting.pdf.

11. Flanders Investment and Trade. 2017a. *Flanders actively supports ecological investments* [Online]. Available: https://www.flandersinvestmentandtrade. com/invest/en/investing-in-flanders/grant-incentives/flanders-actively-supports-ecological-investments.
12. Flanders Investment and Trade. 2017c. *Various tax breaks for R&D investments in Flanders* [Online]. Available: https://www.flandersinvest-mentandtrade.com/invest/en/investing-in-flanders/rd-and-innovation/ flanders-offers-specific-tax-incentives-foreign-businesses-0.
13. Ibid.
14. Flanders Investment and Trade. 2017b. *R&D investment deduction: Flanders supports high-tech innovation* [Online]. Available: https://www. flandersinvestmentandtrade.com/invest/en/investing-in-flanders/rd-and-innovation/tax-incentives-rd-activities/rd-investment-deduction.
15. Flemish Government. 2017b. *Environmental permit* [Online]. Available: http://www.flanders.be/en/enterprise-and-investment/environmental-permit.
16. OVAM. 2017b. *Determine the environmental impact of your product with the ecolizer and make the difference!* [Online]. Available: http://www. ecolizer.be/.
17. OVAM. 2017h. *Integrate sustainability in your innovation process with the OVAM SIS toolkit* [Online]. Available: http://www.ecodesignlink.com/ en/tools/sis-toolkit-1.
18. OVAM. 2014b. Manual OVAM SIS toolkit. Available: http://www. ecodesignlink.com/images/filelib/ManualOVAMSISToolkit_3648.pdf.
19. OVAM. 2017d. *Ecoclusters: promotion of industrial symbiosis* [Online]. Available: https://www.ovam.be/ecoclusters-bevordering-van-industriele-symbiose.
20. Symbiose. 2017. *Industrial symbiosis as a lever for the circular economy* [Online]. Available: http://www.smartsymbiose.com/.
21. OVAM. 2017o. *VLAJO and OVAM reward mini companies* [Online]. Available: https://www.ovam.be/vlajo-en-ovam-belonen-mini-ondernemingen.
22. OVAM. 2017k. *Material scan* [Online]. Available: http://ovam.be/ materialenscan.
23. OVAM. 2017c. *Eco-efficiency scan* [Online]. Available: https://services. ovam.be/ecoscan-extern/views/info/wat.seam;jsessionid=65468B154C7 FB7B04C088331A4845EB0.jb-ex-05-00.

24. OVAM. 2017g. *The GreenVentscan* [Online]. Available: https://www.ovam.be/milieu-gezondheid/duurzame-evenementen/de-groeneventscan.
25. OVAM. 2017f. *The greenVent award* [Online]. Available: https://www.ovam.be/milieu-gezondheid/duurzame-evenementen/de-groenevent-award.
26. OVAM. 2017e. *Ecodesign in higher education? The EHE kit gets you started* [Online]. Available: http://www.ecodesignlink.com/en/tools/ehe-kit.
27. OVAM. 2014a. Ecodesign in higher education guide. Available: http://www.ecodesignlink.com/images/filelib/1EHEkitguide_3528.pdf.
28. OVAM. 2017n. *Tools* [Online]. Available: http://www.ecodesignlink.be/en/tools.
29. Flemish Government. 2017e. *Purchase policy* [Online]. Available: https://overheid.vlaanderen.be/aankoopbeleid.
30. Flemish Government. 2017a. *E-procurement policy* [Online]. Available: https://overheid.vlaanderen.be/beleid-eprocurement.
31. Flemish Government. 2017f. *SME participation* [Online]. Available: https://overheid.vlaanderen.be/kmo-participatie.
32. Flemish Government. 2017c. *Integrity in public procurement* [Online]. Available: https://overheid.vlaanderen.be/integriteit-bij-overheidsopdrachten.
33. Flemish Government. 2017h. *Sustainable and innovative public procurement* [Online]. Available: https://overheid.vlaanderen.be/duurzame-innovatieve-overheidsopdrachten.
34. Flemish Government. 2017d. *Public procurement policy* [Online]. Available: https://overheid.vlaanderen.be/beleid-duurzame-overheidsopdrachten.
35. Flemish Government. 2017g. *Strategic projects* [Online]. Available: https://overheid.vlaanderen.be/strategische-projecten.
36. OVAM. 2017j. *Material conscious building in circuits* [Online]. Available: https://www.ovam.be/afval-materialen/wetgeving-en-beleidsplannen-of-programmas/bouw.
37. OVAM. 2017i. *A management system and unit rules for recycled granulates* [Online]. Available: https://www.ovam.be/afval-materialen/specifieke-afvalstromen-materiaalkringlopen/materiaalbewust-bouwen-in-kringlopen/een-beheersysteem-en-eenheidsreglement-voor-gerecycleerde-granulaten.
38. OVAM. 2017l. *Selectively demolish and dismantle* [Online]. Available: https://www.ovam.be/afval-materialen/specifieke-afvalstromen-materiaalkringlopen/materiaalbewust-bouwen-in-kringlopen/selectief-slopen-ontmantelen.

39. OVAM. 2017a. *The cycles of non-stony materials* [Online]. Available: https://www.ovam.be/afval-materialen/specifieke-afvalstromen-materiaalkringlopen/materiaalbewust-bouwen-in-kringlopen/niet-steenachtige-kringloop.
40. OVAM. 2017m. *Stoney cycle* [Online]. Available: https://www.ovam.be/afval-materialen/specifieke-afvalstromen-materiaalkringlopen/materiaalbewust-bouwen-in-kringlopen/steenachtige-kringloop.
41. Department of Environment. 2017. *What is a green deal?* [Online]. Available: https://www.lne.be/wat-is-een-green-deal.
42. Flanders Circular. 2017a. *Green deal circular buying* [Online].
43. Ibid.
44. Flanders Circular. 2017b. *Masterclass circular economics* [Online]. Available: http://www.vlaanderen-circulair.be/nl/masterclass-circulaire-economie.

References

Department of Environment. 2017. *What is a green deal?* [Online]. Available: https://www.lne.be/wat-is-een-green-deal.

EnergyVille. 2016. *Inventory renewables Flanders 2005–2015* [Online]. Available: http://www.energyville.be/en/nieuwsbericht/inventory-renewables-flanders-2005-2015.

European Commission. 2017. *Flanders* [Online]. Available: https://ec.europa.eu/growth/tools-databases/regional-innovation-monitor/base-profile/flanders.

Flanders Circular. 2017a. *Green deal circular buying* [Online].

———. 2017b. *Masterclass circular economics* [Online]. Available: http://www.vlaanderen-circulair.be/nl/masterclass-circulaire-economie.

Flanders Environment Agency. 2014. Megatrends: Far-reaching, but also out of reach? How do megatrends influence the environment in Flanders? MIRA Future Outlook Report 2014. Available: https://en.vmm.be/news/archive/megatrends-far-reaching-but-also-out-of-reach.

———. 2015. MIRA climate report 2015: About observed and future climate changes in Flanders and Belgium. Available: http://www.milieurapport.be/en/publications/topic-reports/mira-climate-report-2015-about-observed-and-future-climate-changes-in-flanders-and-belgium/.

———. 2017a. Air quality in the Flemish region 2015. Available: https://en. vmm.be/publications/air-quality-in-the-flemish-region-2015.

———. 2017b. MIRA system balance 2017: Environmental challenges for the energy, mobility and food systems in Flanders. Available: http://milieurap-port.be/en/publications/other-reports/mira-system-balance-2017-environmental-challenges-for-the-energy-mobility-and-food-systems.

Flanders Investment and Trade. 2017a. *Flanders actively supports ecological invest-ments* [Online]. Available: https://www.flandersinvestmentandtrade.com/invest/en/investing-in-flanders/grant-incentives/flanders-actively-supports-ecological-investments.

———. 2017b. *R&D investment deduction: Flanders supports high-tech innova-tion* [Online]. Available: https://www.flandersinvestmentandtrade.com/invest/en/investing-in-flanders/rd-and-innovation/tax-incentives-rd-activities/rd-investment-deduction.

———. 2017c. *Various tax breaks for R&D investments in Flanders* [Online]. Available: https://www.flandersinvestmentandtrade.com/invest/en/invest-ing-in-flanders/rd-and-innovation/flanders-offers-specific-tax-incentives-foreign-businesses-0.

———. 2017d. *Why Flanders is a trade powerhouse* [Online]. Available: https://www.flandersinvestmentandtrade.com/invest/en/investing-in-flanders/why-flanders-trade-powerhouse.

Flemish Government. 2017a. *E-procurement policy* [Online]. Available: https://overheid.vlaanderen.be/beleid-eprocurement.

———. 2017b. *Environmental permit* [Online]. Available: http://www.flanders. be/en/enterprise-and-investment/environmental-permit.

———. 2017c. *Integrity in public procurement* [Online]. Available: https://over-heid.vlaanderen.be/integriteit-bij-overheidsopdrachten

———. 2017d. *Public procurement policy* [Online]. Available: https://overheid. vlaanderen.be/beleid-duurzame-overheidsopdrachten

———. 2017e. *Purchase policy* [Online]. Available: https://overheid.vlaanderen. be/aankoopbeleid.

———. 2017f. *SME participation* [Online]. Available: https://overheid.vlaan-deren.be/kmo-participatie

———. 2017g. *Strategic projects* [Online]. Available: https://overheid.vlaan-deren.be/strategische-projecten

———. 2017h. *Sustainable and innovative public procurement* [Online]. Available: https://overheid.vlaanderen.be/duurzame-innovatieve-overheidsopdrachten.

OVAM. 2011. SIS brainstorm poster. Available: http://www.ecodesignlink. com/images/filelib/OVAMSISBrainstormposterEN2011m10d12_5267.pdf.

———. 2014a. Ecodesign in higher education guide. Available: http://www.ecodesignlink.com/images/filelib/1EHEkitguide_3528.pdf.

———. 2014b. Manual OVAM SIS toolkit. Available: http://www.ecodesign-link.com/images/filelib/ManualOVAMSISToolkit_3648.pdf.

———. 2016. Activities report 2016. Available: http://www.ovamenglish.be/sites/default/files/atoms/files/OVAM_activiteitenoverzicht_2016_DRUK_EN_LR_SPREAD.pdf.

———. 2017a. *The cycles of non-stony materials* [Online]. Available: https://www.ovam.be/afval-materialen/specifieke-afvalstromen-materiaalkringlopen/materiaalbewust-bouwen-in-kringlopen/niet-steenachtige-kringloop.

———. 2017b. *Determine the environmental impact of your product with the ecolizer and make the difference!* [Online]. Available: http://www.ecolizer.be/.

———. 2017c. *Eco-efficiency scan* [Online]. Available: https://services.ovam.be/ecoscan-extern/views/info/wat.seam;jsessionid=65468B154C7FB7B04C088331A4845EB0.jb-ex-05-00.

———. 2017d. *Ecoclusters: Promotion of industrial symbiosis* [Online]. Available: https://www.ovam.be/ecoclusters-bevordering-van-industriele-symbiose.

———. 2017e. *Ecodesign in higher education? The EHE kit gets you started* [Online]. Available: http://www.ecodesignlink.com/en/tools/ehe-kit.

———. 2017f. *The greenVent award* [Online]. Available: https://www.ovam.be/milieu-gezondheid/duurzame-evenementen/de-groenevent-award.

———. 2017g. *The GreenVentscan* [Online]. Available: https://www.ovam.be/milieu-gezondheid/duurzame-evenementen/de-groeneventscan.

———. 2017h. *Integrate sustainability in your innovation process with the OVAM SIS toolkit* [Online]. Available: http://www.ecodesignlink.com/en/tools/sis-toolkit-1.

———. 2017i. *A management system and unit rules for recycled granulates* [Online]. Available: https://www.ovam.be/afval-materialen/specifieke-afvalstromen-materiaalkringlopen/materiaalbewust-bouwen-in-kringlopen/een-beheersysteem-en-eenheidsreglement-voor-gerecycleerde-granulaten.

———. 2017j. *Material conscious building in circuits* [Online]. Available: https://www.ovam.be/afval-materialen/wetgeving-en-beleidsplannen-of-programmas/bouw.

———. 2017k. *Material scan* [Online]. Available: http://ovam.be/materialenscan.

———. 2017l. *Selectively demolish and dismantle* [Online]. Available: https://www.ovam.be/afval-materialen/specifieke-afvalstromen-materiaalkringlopen/materiaalbewust-bouwen-in-kringlopen/selectief-slopen-ontmantelen.

———. 2017m. *Stoney cycle* [Online]. Available: https://www.ovam.be/afval-materialen/specifieke-afvalstromen-materiaalkringlopen/materiaalbewust-bouwen-in-kringlopen/steenachtige-kringloop.

———. 2017n. *Tools* [Online]. Available: http://www.ecodesignlink.be/en/tools.

———. 2017o. *VLAJO and OVAM reward mini companies* [Online]. Available: https://www.ovam.be/vlajo-en-ovam-belonen-mini-ondernemingen.

Symbiose. 2017. *Industrial symbiosis as a lever for the circular economy* [Online]. Available: http://www.smartsymbiose.com/.

The Brussels Times. 2016. *Regional economic prospects for 2016–2021 – Slightly lower growth expected in Wallonia than in Flanders* [Online]. Available: http://www.brusselstimes.com/belgium/5964/regional-economic-prospects-for-2016-2021-slightly-lower-growth-expected-in-wallonia-than-in-flanders.

Vlakwa. 2015. *The socio-economic importance of water in Flanders*. Available: http://www.vlakwa.be/fileadmin/media/pdf/20150605_samenvatting.pdf.

6

Natural Resource Management and the Circular Economy in New South Wales

Introduction

New South Wales (NSW) is Australia's largest state economy, with 33 percent of the nation's GDP in 2015–2016. The next largest, Victoria, contributes 22 percent. Services dominate the state economy, accounting for one-third of all NSW exports compared with 17 percent for Australia overall.[1] The NSW economy grew by 3.5 percent in 2015–2016, well above its long-term trend rate of 2.5 percent, while employment expanded by 3.8 percent. Domestic demand continues to be the driving force behind the NSW economy as the nation transitions away from mining investment. In addition, the $73.2 billion state infrastructure investment planned over the coming years will be a driver of economic growth.[2]

Challenges to the Linear Economy

NSW is experiencing a variety of challenges to its traditional linear economy as described below through an assortment of examples.

© The Author(s) 2018
R. C. Brears, *Natural Resource Management and the Circular Economy*,
Palgrave Studies in Natural Resource Management,
https://doi.org/10.1007/978-3-319-71888-0_6

Air Pollution

Most parts of NSW experienced good air quality during 2016 with the Air Quality Index in the 'very good', 'good', or 'fair' category for at least 92 percent of the time in Sydney, 96 percent in the Illawarra and Southwest Slopes regions, and close to 100 percent of the time in all other regions.[3] Nonetheless, air quality can deteriorate during heatwaves, as evidenced in 2017 when NSW Health issued an air pollution alert for Sydney, as well as rural areas, warning that an increased level of ozone in the atmosphere would likely affect residents with respiratory problems.[4]

Climate Change

Average temperatures in NSW are expected to rise by 0.7°C soon (2020–2039), increasing by around 2.1°C in the far future (2060–2079). Heatwaves are projected to occur more often and last longer; up to 3.5 days more on average. By 2030, there is projected to be up to 10 more heatwave days per year and by 2070 up to 33 more in northern NSW, while in the south there will be up to 7 more heatwave days.[5] There may be a decrease in annual rainfall and runoff in inland catchments and minor increases in coastal catchments by 2030. Climate change is also likely to result in an increase in evaporation throughout the catchments, with a 22 percent increase in pan-evaporation in inland catchments and a 9 percent increase in coastal catchments by 2070.[6]

Energy

NSW meets most of its demands for energy from non-renewable sources (around 94 percent in 2012–2016) including coal, gas, and petroleum products. Diversification of NSW electricity supplies is growing strongly with an increase in renewable-based capacity and generation. Electricity generated from renewable sources has increased from 6.1 percent in 2008 to 10.8 percent in 2014. Nonetheless, coal is used to generate over 82 percent of electricity in the state.[7]

Population Growth

In 2014, the population of NSW was 7.52 million, 64 percent of whom were in the Greater Sydney area. By 2031, the population is expected to grow to 9.23 million and reach 9.9 million by 2036, with most of this growth anticipated to be in Sydney. The population of the state will continue to age with significant increases in older age groups. At the same time, there is a growth in younger age groups, particularly children aged 0–4 years and people in their 20s. The combination of rising population, increasing urbanisation, and an ageing population will lead to a greater demand for housing, energy, and water, as well as an increase in the waste generated.[8–10]

Waste

Total waste generation in NSW has continued to rise, with 2012–2013 per person waste in the Metropolitan Levy Area (an amalgamation of the former Sydney Metropolitan Area as well as the Hunter, Central Coast, and Illawarra areas) 22 percent higher than the base year of 2002–2003, while total waste generated across NSW was 42 percent higher in 2012–2013 as compared to the base year. Over this period, the amount of construction and demolition waste and municipal waste generated increased at a substantially higher rate than commercial and industrial waste. This is likely due to the influence of population and economic growth. Overall, it is estimated that the amount of waste generated in NSW over the next 20 years will be 159 percent higher than in 2002.[11,12]

Water

Long-term average water use in NSW is around 7000 gigalitres per year, but use is variable and depends on rainfall and flow conditions. There is high demand for the state's water resources, with urban water use having ongoing impacts on river flow in areas of significant population density.

The amount of water supplied to NSW cities and towns has been increasing by 1.35 percent per annum over the past seven years. Increasing demand is placing stress on the health of natural flow patterns, impacting other water users including agriculture which accounts for up to 80 percent of NSW's water use (depending on conditions, for example, droughts, etc.).[13]

Upstream Fiscal Tools

NSW has implemented a variety of upstream fiscal tools to develop the circular economy and encourage a life cycle perspective to be taken by economic actors in an attempt to decouple economic growth from resource use and associated environmental impacts.

Energy Savings Scheme

The Energy Savings Scheme (ESS) reduces energy consumption in NSW by creating financial incentives for businesses to invest in energy saving projects. Businesses work with voluntary participants, known as Accredited Certificate Providers, to implement projects that can create energy savings certificates for each notional megawatt hour (MWh) of energy the project saves. Energy savings certificates can then be sold to mandatory participants, known as Scheme Participants (electricity retailers, direct suppliers of electricity, and market customers) who have an obligation under the Scheme to meet energy savings targets. Each Scheme Participant must meet an individual energy savings target calculated as a percentage of their liable acquisitions (any purchase of electricity for consumption by, or sale to, end users in NSW, or for use in NSW), with the Scheme's overall energy savings target increasing each year (Table 6.1).[14] Scheme Participants can choose to meet their target by acquiring or surrendering an equivalent number of energy savings certificates, with the price of certificates varying due to supply and demand and therefore can fluctuate depending on market conditions. Alternatively, Scheme Participants can pay the Scheme Penalty Rate. In 2016, the rate was set at $27.03 per notional MWh.

Table 6.1 Energy Savings Scheme targets

Year	Energy savings target
2009*	1.0%
2010	1.5%
2011	2.5%
2012	3.5%
2013	4.5%
2014	5.0%
2015	5.0%
2016	7.0%
2017	7.5%
2018	8.0%
2019–2025	8.5%

*half year from 1 July

In 2017, the penalty rate is $27.48 per notional MWh to reflect an upwards movement in the Consumer Price Index.[15–18]

Upstream Non-Fiscal Tools

NSW has implemented a variety of upstream non-fiscal tools to develop the circular economy and encourage a life cycle perspective to be taken by economic actors in an attempt to decouple economic growth from resource use and associated environmental impacts.

Clean Energy Strategies for Business

The NSW Government has launched the Clean Energy Strategies for Business programme to support businesses in developing a strategy to reach 100 percent renewable energy or an emissions reduction target. Working in collaboration with external advisors, participating businesses will identify opportunities to achieve their targets such as evaluating energy efficiency projects, on-site solar or other renewable energy, off-site procurement of renewable energy or purchasing carbon emission offsets. Each participating business will receive funding of up to $10,000 with participants needing to contribute an additional $5000 to the project. Funding for capital upgrades is not included in the programme.[19]

Sustainability Advantage

The Sustainability Advantage programme, managed by the NSW Office of Environment and Heritage, assists organisations of all types (not for profit, government, and small, medium, or large business) from all sectors (industries, manufacturing, commercial property, registered clubs, health, aged care, transport, education, etc.) across NSW to increase their competitiveness and improve their bottom lines through better environmental practices. While the actual results depend on each participant's own efforts, Sustainability Advantage participants receive expert advice, training, and business tools including practical workshops, a comprehensive range of guides, case studies and templates, one-on-one specialist support, facilitated networking and targeted seminars, and access to an extended network of like-minded organisations.[20]

Sustainability Advantage Recognition Scheme

The NSW Office of Environment and Heritage promotes the environmental achievements of participating organisations through its Sustainability Advantage Recognition Scheme. The scheme acknowledges those organisations that have committed themselves to achieving real environmental improvements through the programme in the areas of:

- *Participation*: Actively participated in the programme
- *Leadership*: Demonstrated leadership and commitment to sustainability
- *Planning*: Established planning and management systems for environmental practice, including processes for continuous improvement
- *Engagement*: Promoted sustainability practices internally among staff and externally with suppliers, customers, and the wider industry
- *Achievements*: Demonstrated environmental outcomes in the last 12 months[21]

To become a member of the Sustainability Advantage programme, organisations need to commit to at least 12 months' participation and make a modest financial contribution, complete an initial sustainability management diagnostic, undertake selected Sustainability Advantage modules, participate in networking meetings and other sustainability events, and share progress reports with the organisers. Regarding the management diagnostic aspect, this assessment will identify potential business benefits that can be gained from more sustainable work practices and provide a clear path of action. Sustainability Advantage will then design achievable programmes specific to each organisation's requirements, based on a series of modules selected from the ones summarised in Table 6.2. Upon joining Sustainability Advantage, members can progress to Sustainability Advantage Bronze, Silver, Gold, and Platinum Partner, with the progress through each level summarised in Table 6.3.

Energy Efficiency Training Courses

The NSW Office of Environment and Heritage offers businesses comprehensive training programmes to help them gain a competitive advantage through reduced energy consumption, improved productivity, and lower operating costs. Examples of courses include:

- *Building the business case*: Participants are taught how to plan, develop, and communicate the business case for energy efficiency projects
- *Introduction to energy management*: A practical and interactive course designed to assist businesses in understanding their energy use, collecting and managing data, and reducing costs through increased efficiency
- *Cogeneration feasibility*: Participants will understand and assess the feasibility of cogeneration and tri-generation for businesses[22]

Downstream Fiscal Tools

NSW has implemented a variety of downstream fiscal tools to develop the circular economy and encourage a life cycle perspective to be taken by economic actors in an attempt to decouple economic growth from resource use and associated environmental impacts.

Table 6.2 Sustainability Advantage modules

Module	Description
Business Planning for Sustainability	This module explores the business case for sustainability and how a firm's culture, purpose, and operations may change. Partners first map their sustainability vision and list their long-term goals. The next session determines what is needed to deliver the goals along with appropriate plans, key performance indicators, and reporting. The organisation then implements the plan for at least 12 months, reviewing it throughout
Resource Productivity	This module helps organisations assess their baseline data on material flows, identify opportunities through a business-cost analysis, execute a resource efficiency action plan, and establish internal processes to drive innovation and savings. This module requires the establishment of a dedicated efficiency team that pursues resource opportunities over 12 months or more
Employee Engagement	This module considers how best to build support for sustainability initiatives through education and training, leadership behaviour, social events, peer and internal forums, communication, signage, mentoring, and recognition
Supply-Chain Management	This module identifies the business case for working with key suppliers and customers to reduce risk and cost and get the best environmental results
Carbon Management	This module looks at the most beneficial strategies for managing and reducing an organisation's carbon footprint as well as how to measure and minimise an organisation's carbon output
Climate Change Risk and Adaptation	Organisations will learn about the potential impact of climate change-related risks, enabling risks and opportunities to be identified
Environmental Risk and Responsibility	This module helps organisations assess their compliance and decide which action is needed to mitigate risk
External Stakeholder Engagement	This module looks at complaint-handling, building new partnerships proactively and effectively through initiatives, and explores how financial markets and non-governmental organisations are calling for more reporting on sustainability performance
Green Lean	This module will assist organisations wishing to streamline key systems and processes, saving time and resources and increasing productivity

Office of Environment and Heritage. 2017e. *Sustainability advantage modules* [Online]. Available: http://www.environment.nsw.gov.au/ sustainabilityadvantage/modules.htm

Table 6.3 Sustainability Advantage Recognition levels

Level	Length of time required	Requirements
Bronze	Member for at least 12 months	Recognises organisations that can demonstrate commitment to business sustainability
Silver	Bronze for at least 12 months	Recognises organisations that can demonstrate significant environmental achievements
Gold	Silver for at least two years	Recognises organisations that can demonstrate outstanding environmental achievement and leadership
Platinum	Gold for at least three years	Recognises an organisation that can demonstrate innovation, performance, and competitive advantage through practices that achieve 'net zero' impact on the environment. Platinum Partners also implement platinum projects that are innovative projects that create a transformational shift in an organisation's business model, product, service, or process

Office of Environment and Heritage. 2014. Sustainability advantage pathways to recognition. Available: http://www.environment.nsw.gov.au/resources/business/sustainabilityadvantage/140609-SA-recognition-pathways.pdf

Waste Less, Recycle More: Business Recycling Grants

NSW's Waste Less, Recycle More initiative to stimulate new investment and transform waste and recycling in the state provides $35 million to help NSW businesses reduce waste and boost recycling. Five key programmes provide funding to support businesses through advice, networking, and equipment that will make it easier to recycle waste rather than send it to the landfill.

Bin Trim Business Grants

The Bin Trim Business Grant funds waste assessments for small and medium-sized businesses with individual grants of between $50,000 and $400,000 given out. The assessors are funded by the NSW Environmental Protection Authority (EPA). They undertake free assessments, produce tailored action plans, and provide support to businesses to reduce waste and increase recycling.[23]

Bin Trim Rebates Program

The Bin Trim Rebates Program aims to increase workplace recycling by providing small and medium-sized businesses with rebates covering up to 50 percent of the cost of installing small-scale, on-site recycling equipment with rebates ranging between $1000 and $50,000.[24]

Circulate, NSW EPA Industrial Ecology Program

The Circulate, NSW EPA Industrial Ecology Program is assisting organisations to achieve competitiveness and improve their bottom lines through better environmental practices. With $4.2 million of funding available, applicants from a variety of sectors, including medium-to-large enterprises, businesses, not-for-profit organisations, and government agencies, can apply for individual grants between $100,000 to $500,000 to have industrial ecology facilitators work with them to help identify and implement projects that will reduce waste or provide them with resources to create new material. Facilitators will help organisations manage different types of waste including timber, plastics, organics, and building operational waste with the programme focusing on the recovery of waste that is currently being sent to the landfill.[25]

Civil Construction Market Pilot Program

The Civil Construction Market Pilot Program promoted the use of waste from one civil construction project as a useful input into another. The programme, which ran until May 2017, aimed to increase resources productivity, reduce the costs of coordination, and minimise the risks involved in reducing the amount of materials being sent to landfill or being stockpiled. To do this, the EPA offered grants ranging from $20,000 to $75,000 to fund consultants, contractors, waste service providers, and local government personnel who could divert construction and demolition waste from landfill through reuse, recycling, and industrial ecology projects in the NSW civil construction sector. Some of the benefits of being part of the programme included

- Opportunities for information sharing, collaboration, and capacity building
- Access to specialist advice
- Branding and marketing support for projects established through the programme
- Monitoring, evaluation, and reporting for projects established through the programme
- Liaison with government consent authorities
- Flow-on bottom line benefits from applying industrial ecology principles to projects
- Opportunities to increase the organisation's commitment to sustainability and innovation[26]

Organics Market Development Grant Program

The Organics Market Development Grant Program provides grant funding of $2.55 million over two years to support the long-term expansion of the market in NSW for recycled organics that have been diverted from landfill. The objectives of the grant are to:

- Facilitate the expansion of the NSW market for recycled organic materials by an additional 70,000 tonnes by June 2017
- Reduce contamination from source-separated organic collection systems to improve end-market confidence in recycled organic products
- Change awareness, knowledge, behaviours, and practices around the use of recycled organic products

  The programme has two streams:

- *Stream 1—Product quality*: Grants of between $10,000 and $50,000 are available for projects that can deliver outcomes to improve the quality of recycled organic products
- *Stream 2—Market development*: Contestable grants of between $50,000 and $500,000 are available for projects that develop new markets or expand existing markets for recycled organics[27]

Recycling Innovation Fund

The Recycling Innovation Fund provided $15 million for projects that created new recycling infrastructure solutions, established or expanded recycling material markets through research and development, and increased the efficiency of recycling facilities for specific targeted wastes. From 2017 to 2021, an additional $5 million will be available through contestable grants for industry, councils, not-for-profit organisations, and charities to further develop projects that provide innovative solutions to targeted waste types in NSW including infrastructure and research.[28]

Container Deposit Scheme

Since 1 July 2017, NSW has been operating its Container Deposit Scheme in which cans and bottles are deposited in exchange for a refund. This helps meet the state's target of reducing litter by 40 percent by 2020. Under the scheme

- People in NSW can return empty prescribed beverage containers between 150 ml and 3 litres to collection points for a 10-cent refund.
- A single Scheme Coordinator is responsible for the financial management of the scheme and for ensuring the scheme meets state-wide targets.
- Beverage suppliers are responsible for covering the costs of refunds through agreements with the Scheme Coordinator.[29]

Waste and Recycling Infrastructure Fund

The Waste and Recycling Infrastructure Fund is designed to accelerate and stimulate investment in infrastructure to increase processing capacity of waste recyclers in waste levy-paying areas of NSW. There are two types of funding programmes available as follows:

Resource Recovery Facility Expansion and Enhancement Program

The Resource Recovery Facility Expansion and Enhancement Program aims to increase recycling of household, industry, and business waste. Councils, industry, business, and not-for-profit organisations who operate existing licensed waste management facilities in NSW and wish to invest in new equipment and upgrades that will boost their recycling processing capacity can apply for individual grants of between $100,000 to $1 million to cover up to 50 percent of project capital costs. Construction and Demolition (C&D) operators can also apply for grants as the C&D waste sector has the potential to return large volumes of recovered material into the economy, reducing the sector's environmental impact.[30,31]

Major Resource Recovery Infrastructure Program

The $43.1 million, three-year Major Resource Recovery Infrastructure Program aims to accelerate and stimulate investment in new waste and recycling infrastructure. Individual grants of between $1 million and $5 million are available for the private sector and not-for-profit organisations and grants of between $1 million and $10 million are available for councils with all grants covering up to 50 percent of project capital costs. Projects are assessed for their capacity to cost-effectively increase household and business waste recycling, with priorities placed on

- Recovery of recyclables from sorted and unsorted waste from businesses and households
- Reuse, recycling, and reprocessing of recyclable materials from businesses and households such as plastics, timber, paper, cardboard, consumer packaging, and tyres
- Processing, stabilisation, and energy recovery from residual business and household waste

All applications are required to submit a business case detailing their project's feasibility. The EPA has commissioned a consultancy firm to provide free support to applicants to help prepare their business case, with all grant applicants able to obtain the use of the consultancy's proprietary Cost Benefit Analysis tools. In addition, the EPA has established a panel of expert contractors to provide free advice on infrastructure and procurement. Furthermore, on-request advisory services are available to council and private grant recipients in many areas including

- *Procurement/economic/financial reviews*: Peer reviews of infrastructure planning approaches or critical reviews of project timelines and risk plans
- *Planning/licensing*: Guidance on environmental protection licensing or development application procedures
- *Technical/engineering*: Review of the design of infrastructure and civil works or advice on equipment purchasing
- *Compliance*: Ensuring compliance with legislative and policy requirements
- *General/other*: Overcoming challenging problems that require additional assistance, or which do not fall within the categories above[32]

Downstream Non-Fiscal Tools

NSW has implemented a variety of downstream non-fiscal tools to develop the circular economy and encourage a life cycle perspective to be taken by economic actors in an attempt to decouple economic growth from resource use and associated environmental impacts.

Australian Packaging Covenant

The Australian Packaging Covenant, which brings government, industry, and community groups together to fund projects that address packaging sustainability issues, jointly co-funded NSW projects to increase expanded polystyrene (EPS) recycling rates as the state only recycled less than 10 percent of its EPS in 2011.[33]

Better Practice Guidelines for Waste Management and Recycling in Commercial and Industrial Facilities

The Better Practice Guidelines for Waste Management and Recycling in Commercial and Industrial Facilities, published by the EPA for facility managers, property managers, commercial tenants, property developers, building designers, council officers, and waste and recycling providers, offers guidance on how owners and managers of commercial and industrial premises can achieve major cost savings and environmental benefits from optimising their waste and recycling systems. The guidelines cover design of waste management systems and day-to-day operations for premises including office buildings, retail outlets, group retail centres, and hospitality and accommodation facilities. The guidelines provide checklists, specifications, case studies as well as practical advice including

- Opportunities to avoid waste
- Ways to increase the yield and quality of recyclable materials
- Equipment choices and layout
- Spatial requirements for storage, handling, and collection of waste
- Information about contracts for waste services[34]

Government Resource Efficiency Policy

The Government Resource Efficiency Policy (GREP) aims to reduce the operating costs of NSW Government agencies and ensure they provide leadership in resource productivity. Specifically, GREP requires government agencies to

- Meet the challenge of rising costs for energy, water, clean air, and waste management with GREP having a range of targets in each area (summarised in Table 6.4)
- Use purchasing power to drive down the costs of resource-efficient technologies and services

Table 6.4 Government Resource Efficiency Policy targets

Measure	Targets	Minimum standards
Energy	Targets to Undertake Energy Efficiency Projects	All government-sector clusters will undertake energy efficiency projects at sites representing 90 percent of their billed energy use by the end of 2023–2024, with an interim target of 55 percent for health and 40 percent for other clusters by the end of 2017–2018
	Minimum National Australian Built Environment Rating System (NABERS) Energy Ratings for Offices and Data Centres	Large owned and leased office buildings will have achieved and maintained a NABERS Energy rating of at least 4.5 stars by June 2017. All data centres will have achieved a minimum infrastructure and IT equipment NABERS Energy rating of 4.5 stars by June 2017
	Minimum Standards for New Electrical Appliances and Equipment	All new electrical equipment purchased by government must be at least the market average star rating. In categories where no star ratings are available, equipment purchased should be recognised as high efficiency either by being ENERGY STAR® accredited, in a high-efficiency band under Australian Standards or being above-average efficiency of Greenhouse and Energy Minimum Standards (GEMS) registered products
	Minimum Standards for New Buildings	All new office buildings and fit-outs will be designed and built to a predicted performance of at least 4.5 stars NABERS Energy rating. For other building types, new facilities with project costs over $10 million should be designed and built so that energy consumption is predicted to be at least 10 percent lower than if built to minimum compliance with National Construction Code requirements
	Identify and Enable Solar Leasing Opportunities	Small government sites had conducted self-assessments for their suitability for solar leasing by July 2015
	Minimum Fuel Efficiency Standards for New Light Vehicles	Improve minimum fuel efficiency standards for new light vehicles so that the average NSW Government purchase is at least the market average fuel efficiency by vehicle category
	Purchase 6 percent GreenPower	Purchase a minimum of 6 percent GreenPower

(continued)

Table 6.4 (continued)

Measure	Targets	Minimum standards
Water	Report on Water Use	All agencies will report on water use
	Minimum Water Standards for Office Buildings	All new and refurbished owned office buildings and leased office buildings with a net lettable area of over 2000 m² will achieve a whole building NABERS Water rating of 4 stars where cost-effective.
	Minimum standards for New Water-Using Appliances	All new water-using appliances, shower heads, taps, and toilets purchased by agencies must be at least the average Water Efficiency Labelling Scheme star rating by product type
Waste	Report on Top Three Waste Streams	All agencies will report on their top three waste streams by total volume and by total cost
Clean Air	Air Emission Standards for Mobile Non-Road Diesel Plant and Equipment	Contractor-supplied and government-purchased equipment will comply with European Union or United States Environmental Protection Agency standards
	Low-Volatile Organic Compound Surface Coatings	All surface coatings will comply with the Australian Paint Approval Scheme where fit for purpose

NSW Government. 2014. NSW Government resource efficiency policy. Available: http://www.environment.nsw.gov.au/resources/government/NSW-GREP-140567.pdf

Office of Environment and Heritage. 2017a. *Government resource efficiency policy (GREP)* [Online]. Available: http://www.environment.nsw.gov.au/government/policy.htm

- Show leadership by incorporating resource efficiency into decision-making
- Publish annual statements of their performance against the policy

Each year, general government-sector agencies need to submit a GREP progress report to the NSW Office of Environment and Heritage. This is done through an online Centralised Analysis System for Performance of Energy and Resources web portal, which enables agencies to

- Upload off-contract resource use data
- Upload annual GREP statements of compliance
- Report on achievements and activities beyond compliance
- Identify hotspots and opportunities for improvement
- Automatically generate the agency's annual GREP report[35,36]

Environmental Upgrade Agreements

The Local Government Amendment (Environmental Upgrade Agreements) Act of 2010 authorises councils to enter into Environmental Upgrade Agreements (EUAs) with building owners and finance providers. An EUA is an agreement where

- A building owner agrees to carry out upgrade works to improve energy, water, or environmental efficiency or sustainability of the building;
- A finance provider agrees to advance funds to the building owner to finance those environmental upgrades; and
- The council agrees to levy a charge on the relevant land for repaying the advance to the finance provider

This agreement enables the council to levy an environmental upgrade charge on the land in accordance with an agreed repayment schedule. The council then collects the environmental upgrade charge from the building owner and passes it onto the finance provider to repay the funds advanced. The details of the retrofit activity and the total funds advanced are established by the finance provider and property owner and specified in the EUA.[37] Overall, the agreement enables building upgrades to deliver significant savings that can be used to repay the upgrade finance. In addition, owners and tenants can both gain from maximising a building's energy efficiency with EUAs allowing owners and tenants to share the benefits of lower utility bills. To date, tenants have contributed around 50 percent of the cost of EUA-funded energy efficiency projects. One of the key benefits of EUAs is that it is secure finance, so it can be locked in for a longer term, up to 15 years. This increases the likelihood that the savings generated will be greater than the EUA repayments and better match the life of the upgraded equipment. Furthermore, EUA finance is attached to the land. If the land is sold, the EUA may be transferred to the new owners or discharged on settlement. Eligible works that EUAs can fund include

- Energy and water efficiency
- Renewable energy
- Reducing greenhouse gas emissions
- Preventing or reducing pollution
- Reducing the use of materials
- Recovery or recycling of materials
- Monitoring environmental performance
- Encouraging alternatives to car travel, for example, walking and cycling[38]

NSW Renewable Energy Map

The NSW Renewable Energy Map is an interactive online platform that shows users the renewable energy potential of NSW from sunlight, wind, waves, biomass, water, and heat from the Earth. Specifically, the map uses geographic information systems (GIS) software that enables users to discover the state's resource potential and existing infrastructure for solar, wind, geothermal, bioenergy, hydro, and ocean resources.[39]

Environmental Management Systems Guidelines

The NSW Government Environmental Management System Guidelines were developed to facilitate the achievement of improved environmental performance by the construction industry. Contractors seeking to work on major projects (all projects $10 million or more and projects under $10 million if they are environmentally sensitive) will need to have in place a corporate environmental management system accredited with the NSW Government. Contractors that have been subject to any environmental prosecutions or penalties in the preceding three years will also have to demonstrate, by audit, management review or submission of corrective action and system change information, that any shortcomings in their system have been effectively remedied.[40]

The Recyculator

The Recyculator is an interactive online tool that calculates the environmental benefits of large-scale recycling initiatives across 21 material types in NSW. It aims to assist NSW councils, industry, and businesses to estimate the full environmental benefits of recycling, help broaden resource recovery programmes, and communicate these benefits in an understandable way. It measures the greenhouse gas benefits, energy savings, water savings, and landfill space saved, and displays these results. Specifically, users can find out how much their organisation has saved:

- Greenhouse gas abatement by amount of overall carbon dioxide equivalence (CO_2eq) shown in the number of cars permanently removed off the road
- Energy conservation by megajoules (MJ) per household per year translated into the average household energy usage saved
- Water conservation by kilolitres (kL) converted into Olympic-sized swimming pools of water saved
- Landfill space savings by cubic metres (m^3) and converted into the number of 240-litre mobile garbage bins (wheelie bins)[41]

Case Study Summary

NSW is Australia's largest state economy and has experienced above-average economic growth due to strong domestic demand and infrastructure investment. Nonetheless, NSW is experiencing a variety of challenges to its linear economy. Air pollution can deteriorate significantly during heatwaves leading to respiratory problems for residents. Heatwaves will become more intense with climate change predicted to result in longer and more frequent hot days. At the same time, the state will experience water scarcity in inland as well as coastal catchments. NSW produces most of its electricity from coal with renewable energy sources increasing moderately. Over the coming decades, the state will experience high population growth that will increase demand for energy and water as well as significantly increase the amount of waste generated. The increased demand for water will place stress on the health of natural flow patterns, impacting a variety of water users.

Upstream

NSW has implemented a variety of upstream fiscal and non-fiscal tools to develop the circular economy and encourage a life cycle perspective to be taken by economic actors in an attempt to decouple economic growth from resource use and associated environmental impacts. Some of the main tools are as follows:

- NSW's ESS offers financial incentives for businesses to voluntarily invest in energy saving projects with participating businesses receiving certificates for energy saved. These certificates can be sold to mandatory participants including electricity retailers and direct suppliers of electricity.
- The NSW Government has launched the Clean Energy Strategies for Business programme in which advisors will work with participating businesses to identify opportunities to achieve clean energy targets including developing on-site renewable energy or off-site procurement of renewable energy.
- NSW's Sustainability Advantage programme assists all types of organisations across all sectors to increase their bottom-lines through better environmental practices. To help facilitate the success of environmental initiatives, participants receive expert advice, training, guides, case studies, one-on-one support, as well as networking opportunities. Organisations are recognised for real progress made with participant members able to progress from bronze through to gold based on actions undertaken.

- NSW's Energy Efficiency Training Courses programme offers businesses comprehensive training on gaining a competitive advantage through energy efficiency.

Downstream

NSW has implemented a variety of downstream fiscal and non-fiscal tools to develop the circular economy and encourage a life cycle perspective to be taken by economic actors in an attempt to decouple economic growth from resource use and associated environmental impacts. Some of the main tools are as follows:

- NSW's Waste Less, Recycle More Business Recycling Grants initiative funds NSW businesses to reduce waste and increase their recycling with grants available for businesses to carry out waste assessments, install on-site small-scale recycling equipment, work with industrial ecology facilitators to identify projects that will reduce waste or provide resources to create new material, divert construction and demolition waste from landfill through reuse and recycling, and expand the NSW recycled organics market.
- The state's Recycling Innovation Fund is providing opportunities for organisations to develop projects that will create new recycling infrastructure solutions, establish or expand recycling material markets, and increase the efficiency of recycling facilities for specific targeted wastes.
- NSW has begun operating a Container Deposit Scheme in which consumers can return prescribed beverage containers in exchange for a refund.
- NSW's Waste and Recycling Infrastructure Fund operates the Resource Recovery Facility Expansion and Enhancement Program to increase recycling of household, industry, and business waste including C&D waste in addition to a major resource recovery infrastructure programme to encourage investment in new waste and recycling infrastructure.
- The Australian Packaging Covenant, which brings together the public and private sectors to fund projects that address packaging sustainability-related issues, has funded projects in NSW to increase packaging recycling rates.

- The EPA has developed best practice guidelines for commercial and industrial facilities to avoid waste and increase the yield and quality of recyclable materials. The guidelines provide a series of checklists, specifications, case studies, as well as practical information including on how to avoid waste and increase the yield and quality of recyclable material.
- NSW's GREP aims to show government leadership in resource productivity with the policy requiring government agencies to set resource consumption targets, use their purchasing power to drive down the cost of resource-efficient technologies, show leadership by incorporating resource efficiency in decision-making, and publish annual performance reports.
- NSW's EUAs involves building owners conducting resource efficiency upgrades in their buildings, finance providers making finance available for building owners to complete their environmental upgrades, and councils levying a charge for the environmental upgrades with the collected levy passed onto finance providers to repay the funds advanced.
- NSW's Renewable Energy Map is an interactive online platform that shows the renewable energy potential in NSW from a variety of renewable energy sources using GIS software.
- The NSW Environmental Management System Guidelines was developed to improve the construction industry's environmental performance. Large-scale construction projects will need to have in place an accredited environmental management system with failure to comply resulting in prosecution or penalties.
- NSW's Recyculator is an interactive online tool that lets users calculate the environmental benefits of large-scale recycling initiatives across a variety of material types in NSW. The tool assists a variety of stakeholders to estimate the full environmental benefits of recycling, help broaden resource recovery programmes, and communicate the benefits in an understandable way.

Overall, NSW has implemented a variety of upstream and downstream fiscal and non-fiscal tools to develop the circular economy and encourage a life cycle perspective to be taken by economic actors in an attempt to decouple economic growth from resource use and associated environmental impacts. These tools are summarised in Table 6.5.

Table 6.5 NSW case study summary

Tool	Tool type	Tool title	Description	Upstream/Downstream
Fiscal	Environmental Taxes and Charges	Container Deposit Scheme	Consumers can return prescribed beverage containers in exchange for a refund	Downstream
	Subsidies and Incentives	Bin Trim Business Grant	The grant funds organisations to carry out waste assessments for small and medium-sized businesses	Downstream
		Bin Trim Rebates Program	Aims to increase workplace recycling by providing small and medium-sized businesses with rebates to install small-scale, on-site recycling equipment	Downstream
		Circulate, NSW EPA Industrial Ecology Program	Assists organisations to achieve competitiveness and improve their bottom lines through better environmental practices	Downstream
		Civil Construction Market Pilot Program	Promoted the use of waste from one civil construction project as a useful input into another	Downstream
		Organics Market Development Grant Program	Grant funding to support the long-term expansion of the NSW recycled organics market	Downstream
		Recycling Innovation Fund	Funds projects that create new recycling infrastructure solutions, establish, expand recycling material markets or increase the efficiency of recycling facilities for specific wastes	Downstream
		Resource Recovery Facility Expansion and Enhancement Program	Aims to increase recycling of household, industry, and business waste including C&D waste	Downstream
		Major Resource Recovery Infrastructure Program	Encourages investment in new waste and recycling infrastructure	Downstream
	Tradeable Permits	ESS	Businesses voluntarily invest in energy saving projects with certificates generated that can be sold to mandatory participants	Upstream

(continued)

Table 6.5 (continued)

Tool	Tool type	Tool title	Description	Upstream/Downstream
Non-Fiscal	Green Public Procurement	GREP	Requires government agencies to set resource consumption targets, use their purchasing power to drive down the cost of resource-efficient technologies, and show circular economy leadership	Downstream
	Education and Training	Energy Efficiency Training Courses	Businesses can receive comprehensive training on how to gain a competitive advantage through energy efficiency	Upstream
	Raising Industry Awareness and Capacity	Clean Energy Strategies for Business	Advisors will work with participating businesses to identify opportunities to achieve clean energy targets	Upstream
		Better Practice Guidelines for Waste Management and Recycling in Commercial and Industrial Facilities	Guidelines for commercial and industrial facilities to avoid waste and increase the yield and quality of recyclable materials	Downstream
	Industry-Based Standards	Australian Packaging Covenant	Brings together the public and private sectors to fund projects that address packaging sustainability-related issues	Downstream
	Voluntary Agreements	EUAs	Enables councils to enter into agreements with building owners and finance providers, encouraging building owners to conduct resource efficiency upgrades in their buildings	Downstream
	Greening the Supply Chain	Environmental Management System Guidelines	Large-scale construction projects will need to have in place an accredited environmental management system	Downstream
	Environmental Recognition Awards	Sustainability Advantage Recognition Scheme	Assists all types of organisations across all sectors to increase their bottom-lines through better environmental practices. Organisations are also recognised for real progress made	Upstream
	Information-Based Tools	NSW Renewable Energy Map	An interactive online platform that shows the renewable energy potential in NSW	Downstream
		The Recyculator	An interactive online tool that calculates the environmental benefits of large-scale recycling initiatives across a variety of material types in NSW	Downstream

Notes

1. Department of Industry. 2017. *Size of NSW economy* [Online]. Available: https://www.industry.nsw.gov.au/invest-in-nsw/about-nsw/economic-growth/Size-of-NSW-economy.
2. NSW Treasury. 2017. *Economic outlook* [Online]. Available: https://www.treasury.nsw.gov.au/nsw-economy/about-nsw-economy/economic-outlook.
3. Office of Environment and Heritage. 2017f. Towards cleaner air. NSW air quality statement 2016.
4. Sydney Morning Herald. 2017. *Sydney weather: Air pollution alert as temperatures to hit 47 degrees across NSW this week* [Online]. Available: http://www.smh.com.au/environment/weather/sydney-weather-air-pollution-alert-as-temperatures-to-hit-47-across-nsw-this-week-20170109-gtojlj.html.
5. Office of Environment and Heritage. 2016a. Climate change in NSW fact sheet. Available: http://www.environment.nsw.gov.au/research-and-publications/publications-search/climate-change-fact-sheet-climate-change-in-nsw.
6. Department of the Environment and Energy, A. G. 2017. *Climate change impacts in New South Wales* [Online]. Available: https://www.environment.gov.au/climate-change/climate-science/impacts/nsw.
7. NSW EPA. 2015. NSW state of the environment 2015. Available: http://www.epa.nsw.gov.au/soe/soe2015/index.htm.
8. NSW Government. 2016b. Waste less, recyle more. Available: http://www.epa.nsw.gov.au/resources/waste/waste-less-recycle-more-2017-21-brochure-160538.pdf.
9. ID Consulting. 2017. *New South Wales: The return of the premier state?* [Online]. Available: http://blog.id.com.au/2017/population-forecasting/new-south-wales-the-return-of-the-premier-state/.
10. NSW EPA. 2015. NSW state of the environment 2015. Available: http://www.epa.nsw.gov.au/soe/soe2015/index.htm.
11. Ibid.
12. NSW Government. 2016b. Waste less, recyle more. Available: http://www.epa.nsw.gov.au/resources/waste/waste-less-recycle-more-2017-21-brochure-160538.pdf.
13. NSW EPA. 2015. NSW state of the environment 2015. Available: http://www.epa.nsw.gov.au/soe/soe2015/index.htm.

14. Energy Saving Scheme. 2017d. *Targets and penalties* [Online]. Available: http://www.ess.nsw.gov.au/Scheme_Participants/Targets_ and_penalties.
15. Energy Saving Scheme. 2017a. *How the scheme works* [Online]. Available: http://www.ess.nsw.gov.au/How_the_scheme_works.
16. Energy Saving Scheme. 2017c. *Scheme-specific questions* [Online]. Available: http://www.ess.nsw.gov.au/Common_questions/Scheme-specific_ questions.
17. Energy Saving Scheme. 2017d. *Targets and penalties* [Online]. Available: http://www.ess.nsw.gov.au/Scheme_Participants/Targets_ and_penalties.
18. Energy Saving Scheme. 2017b. *Overview of the scheme* [Online]. Available: http://www.ess.nsw.gov.au/How_the_scheme_works/Overview_of_ the_scheme.
19. NSW Planning and Environment. 2017a. *Clean energy strategies for business* [Online]. Available: http://www.resourcesandenergy.nsw.gov.au/energy-consumers/sustainable-energy/clean-energy-strategies-for-business.
20. Office of Environment and Heritage. 2017d. *Sustainability advantage* [Online]. Available: http://www.environment.nsw.gov.au/sustainability advantage/.
21. Office of Environment and Heritage. 2017b. *Recognition scheme* [Online]. Available: http://www.environment.nsw.gov.au/sustainability-advantage/recognition.htm.
22. Office of Environment and Heritage. 2016b. Energy efficiency training courses. Available: http://www.environment.nsw.gov.au/resources/ business/0477-OEH-training-brochure.pdf.
23. NSW EPA. 2017b. *Bin trim business grants program* [Online]. Available: http://www.epa.nsw.gov.au/wastegrants/bin-trim-business.htm.
24. NSW EPA. 2017c. *Bin trim rebates program* [Online]. Available: http:// www.epa.nsw.gov.au/wastegrants/bin-trim-rebates.htm.
25. NSW EPA. 2017d. *Circulate, NSW EPA industrial ecology program* [Online]. Available: http://www.epa.nsw.gov.au/managewaste/indus-trial-ecology.htm.
26. NSW EPA. 2017e. *Civil construction market pilot program* [Online]. Available: http://www.epa.nsw.gov.au/wastegrants/circulate-civil-con-struction-market-program.htm.
27. NSW EPA. 2017g. *Organics market development grants* [Online]. Available: http://www.epa.nsw.gov.au/wastegrants/organics-market.htm.

28. NSW Government. 2016b. Waste less, recyle more. Available: http://www.epa.nsw.gov.au/resources/waste/waste-less-recycle-more-2017-21-brochure-160538.pdf.
29. NSW Government. 2016a. *Can we do it? Yes we can* [Online]. Available: http://www.environment.nsw.gov.au/resources/MinMedia/EPAMinMedia16101901.pdf.
30. NSW EPA. 2017i. *Resource recovery facility expansion and enhancement program* [Online]. Available: http://www.epa.nsw.gov.au/wastegrants/facility-expansion.htm.
31. Office of Environment and Heritage. 2017c. *Resource recovery facility expansion and enhancement* [Online]. Available: http://www.environment.nsw.gov.au/grants/resourcerec.htm.
32. NSW EPA. 2017f. *Infrastructure advisory services* [Online]. Available: http://www.epa.nsw.gov.au/wastegrants/infrastructure-advice.htm.
33. NSW EPA. 2017a. *Australian packaging covenant* [Online]. Available: http://www.epa.nsw.gov.au/wastegrants/packaging-covenant.htm.
34. NSW EPA. 2012. Waste and recycling systems for commercial and industrial premises. Available: http://www.epa.nsw.gov.au/managewaste/commercial-industrial.htm.
35. NSW Government. 2014. NSW Government resource efficiency policy. Available: http://www.environment.nsw.gov.au/resources/government/NSW-GREP-140567.pdf.
36. Office of Environment and Heritage. 2017a. *Government resource efficiency policy (GREP)* [Online]. Available: http://www.environment.nsw.gov.au/government/policy.htm.
37. NSW Government. 2011. Guidelines for environmental upgrade agreements. Available: http://www.environment.nsw.gov.au/resources/business/EUAGuidelines.pdf.
38. Office of Environment and Heritage. 2017g. *Upgrade your building* [Online]. Available: http://www.environment.nsw.gov.au/business/upgrade-agreements.htm.
39. NSW Planning and Environment. 2017b. *NSW renewable energy resources map* [Online]. Available: http://www.resourcesandenergy.nsw.gov.au/investors/renewable-energy/renewable-resources-map.
40. NSW Government. 2017. *Environmental management systems* [Online]. Available: https://www.procurepoint.nsw.gov.au/environmental-management-systems.
41. NSW EPA. 2017h. *The recyculator* [Online]. Available: http://www.epa.nsw.gov.au/warrlocal/recyculator.htm.

References

Department of Industry. 2017. *Size of NSW economy* [Online]. Available: https://www.industry.nsw.gov.au/invest-in-nsw/about-nsw/economic-growth/Size-of-NSW-economy.

Department of the Environment and Energy, A. G. 2017. *Climate change impacts in New South Wales* [Online]. Available: https://www.environment.gov.au/climate-change/climate-science/impacts/nsw.

Energy Saving Scheme. 2017a. *How the scheme works* [Online]. Available: http://www.ess.nsw.gov.au/How_the_scheme_works.

———. 2017b. *Overview of the scheme* [Online]. Available: http://www.ess.nsw.gov.au/How_the_scheme_works/Overview_of_the_scheme.

———. 2017c. *Scheme-specific questions* [Online]. Available: http://www.ess.nsw.gov.au/Common_questions/Scheme-specific_questions.

———. 2017d. *Targets and penalties* [Online]. Available: http://www.ess.nsw.gov.au/Scheme_Participants/Targets_and_penalties.

ID Consulting. 2017. *New South Wales: The return of the premier state?* [Online]. Available: http://blog.id.com.au/2017/population-forecasting/new-south-wales-the-return-of-the-premier-state/.

NSW EPA. 2012. Waste and recycling systems for commercial and industrial premises. Available: http://www.epa.nsw.gov.au/managewaste/commercial-industrial.htm.

———. 2015. NSW state of the environment 2015. Available: http://www.epa.nsw.gov.au/soe/soe2015/index.htm.

———. 2017a. *Australian packaging covenant* [Online]. Available: http://www.epa.nsw.gov.au/wastegrants/packaging-covenant.htm.

———. 2017b. *Bin trim business grants program* [Online]. Available: http://www.epa.nsw.gov.au/wastegrants/bin-trim-business.htm.

———. 2017c. *Bin trim rebates program* [Online]. Available: http://www.epa.nsw.gov.au/wastegrants/bin-trim-rebates.htm.

———. 2017d. *Circulate, NSW EPA industrial ecology program* [Online]. Available: http://www.epa.nsw.gov.au/managewaste/industrial-ecology.htm.

———. 2017e. *Civil construction market pilot program* [Online]. Available: http://www.epa.nsw.gov.au/wastegrants/circulate-civil-construction-market-program.htm.

———. 2017f. *Infrastructure advisory services* [Online]. Available: http://www.epa.nsw.gov.au/wastegrants/infrastructure-advice.htm.

———. 2017g. *Organics market development grants* [Online]. Available: http://www.epa.nsw.gov.au/wastegrants/organics-market.htm.

———. 2017h. *The recyculator* [Online]. Available: http://www.epa.nsw.gov.au/warrlocal/recyculator.htm.

———. 2017i. *Resource recovery facility expansion and enhancement program* [Online]. Available: http://www.epa.nsw.gov.au/wastegrants/facility-expansion.htm.

NSW Government. 2011. Guidelines for environmental upgrade agreements. Available: http://www.environment.nsw.gov.au/resources/business/EUA Guidelines.pdf.

———. 2014. NSW Government resource efficiency policy. Available: http://www.environment.nsw.gov.au/resources/government/NSW-GREP-140567.pdf.

———. 2016a. *Can we do it? Yes we can* [Online]. Available: http://www.environment.nsw.gov.au/resources/MinMedia/EPAMinMedia16101901.pdf.

———. 2016b. Waste less, recycle more. Available: http://www.epa.nsw.gov.au/resources/waste/waste-less-recycle-more-2017-21-brochure-160538.pdf.

———. 2017. *Environmental management systems* [Online]. Available: https://www.procurepoint.nsw.gov.au/environmental-management-systems.

NSW Planning and Environment. 2017a. *Clean energy strategies for business* [Online]. Available: http://www.resourcesandenergy.nsw.gov.au/energy-consumers/sustainable-energy/clean-energy-strategies-for-business.

———. 2017b. *NSW renewable energy resources map* [Online]. Available: http://www.resourcesandenergy.nsw.gov.au/investors/renewable-energy/renewable-resources-map.

NSW Treasury. 2017. *Economic outlook* [Online]. Available: https://www.treasury.nsw.gov.au/nsw-economy/about-nsw-economy/economic-outlook.

Office of Environment and Heritage. 2014. Sustainability advantage pathways to recognition. Available: http://www.environment.nsw.gov.au/resources/business/sustainabilityadvantage/140609-SA-recognition-pathways.pdf.

———. 2016a. Climate change in NSW fact sheet. Available: http://www.environment.nsw.gov.au/research-and-publications/publications-search/climate-change-fact-sheet-climate-change-in-nsw.

———. 2016b. Energy efficiency training courses. Available: http://www.environment.nsw.gov.au/resources/business/0477-OEH-training-brochure.pdf.

———. 2017a. *Government resource efficiency policy (GREP)* [Online]. Available: http://www.environment.nsw.gov.au/government/policy.htm.

———. 2017b. *Recognition scheme* [Online]. Available: http://www.environment.nsw.gov.au/sustainabilityadvantage/recognition.htm.

———. 2017c. *Resource recovery facility expansion and enhancement* [Online]. Available: http://www.environment.nsw.gov.au/grants/resourcerec.htm.

———. 2017d. *Sustainability advantage* [Online]. Available: http://www.environment.nsw.gov.au/sustainabilityadvantage/.

———. 2017e. *Sustainability advantage modules* [Online]. Available: http://www.environment.nsw.gov.au/sustainabilityadvantage/modules.htm.

———. 2017f. *Towards cleaner air*. NSW air quality statement 2016.

———. 2017g. *Upgrade your building* [Online]. Available: http://www.environment.nsw.gov.au/business/upgrade-agreements.htm.

Sydney Morning Herald. 2017. *Sydney weather: Air pollution alert as temperatures to hit 47 degrees across NSW this week* [Online]. Available: http://www.smh.com.au/environment/weather/sydney-weather-air-pollution-alert-as-temperatures-to-hit-47-across-nsw-this-week-20170109-gtojlj.html.

7

Natural Resource Management and the Circular Economy in Denmark

Introduction

Denmark has one of the strongest economies in Europe, with a balanced state budget, stable currency, low-interest rates, and low inflation. Economic growth is projected to be 1.8 percent in 2018 due to increased private consumption and stronger foreign demand for Danish goods and services. The Danish economy is small, open, and focused on trade with the country's main trading partners in Europe being Germany, Sweden, the UK, and Norway. Outside of Europe, the United States and Japan are also important trading partners.[1,2,3] The main economic sectors of Denmark are agriculture, with the country producing enough foodstuffs for around 15 million people, followed by engineering, industrial production, and energy, with the country being the third-largest oil producer in Western Europe.[4]

Challenges to the Linear Economy

Denmark is experiencing a variety of challenges to its traditional linear economy as described below through an assortment of examples.

© The Author(s) 2018
R. C. Brears, *Natural Resource Management and the Circular Economy*,
Palgrave Studies in Natural Resource Management,
https://doi.org/10.1007/978-3-319-71888-0_7

Air Pollution

In Denmark, the largest contributing sectors to air-related emissions are energy use and supply excluding transport (51 percent), followed by road transport (38 percent). While nitrogen oxide emissions have decreased between 1990 and 2010, as have carbon monoxide levels, over 25 percent of the urban population is exposed to PM10 concentrations above the European Union (EU) air quality objectives.[5]

Climate Change

Climate change is likely to result in increased precipitation, milder winters, warmer summers, and more extreme weather events. There will likely be a 6 percent increase in heavy precipitation events between now and 2050; mean winter and summer temperature by 2100 will be 3.5°C and 2.2°C higher, respectively, relative to the period 1961–1990; the number of days with more than 10 mm of precipitation will increase by 7 by 2050; and the number of heatwave days (three consecutive days above 28°C) per year will increase by 1.5 by 2050.[6]

Energy

For many years now there has been a downward trend in consumption of fossil fuels; a trend that is expected to continue up to 2020. After 2020, however, it is projected that consumption of fossil fuels will increase due to a decline in energy improvements in energy consumption, increased demand for electricity from data centres, and a halt in the installation of new wind power capacity due to the discontinuation of subsidies for renewables because of the expiry of EU approvals. This will see the trend of fossil fuel consumption increasing from 450 PJ in 2020 to 520 PJ in 2030. This increase will be met by a rise in coal-based electricity generation, while consumption of oil and natural gas is projected to stay at a relatively constant level after 2020. The renewable share of final energy consumption is likely to reach 40 percent in 2020; however, after 2020, the share will be constant due to no more power stations expected to be converted to biomass, no more offshore wind farms being constructed, and no wind turbines established onshore due to the discontinuation of state subsidies.[7]

Population Growth

Denmark has a population of 5.7 million inhabitants. The country's population is projected to rise to 6 million in 2030, 6.3 million in 2050, and 6.5 million in 2060.[8] One of the largest population centres experiencing growth will be Copenhagen, with the city's population expected to increase by 36 percent over the period 2005–2025.[9]

Waste

Total Danish waste generated, including recycling, incineration, landfilling, temporary storage, and special treatment, was 11.74 million tonnes in 2014, up from 10.99 million tonnes in 2012. The main reason was an increase in quantities of construction and demolition waste caused primarily by higher demolition activity. Meanwhile, over the period 1995–2015, the amount of municipal waste generated per capita increased by 2.1 percent. Nonetheless, since 2012, the share of total waste going to recycling has increased by 2 percent, due to an increase in quantities of household waste and construction waste going to recycling.[10,11]

Water

Since the 1970s, one of the largest challenges for water resources planning and administration has been to regulate the abstraction of surface and groundwater to an acceptable level. The other challenge to water supply has been the pollution of well fields with nitrate from farming, chemicals from old waste dumps and oil tanks, toxic materials from enterprises, and pesticides from urban areas and farmlands contaminating groundwater supplies. Over the period 1991–2005, more than 1300 wells were closed as water supply abstraction wells due to contamination. Today this trend continues with around 100 wells closed each year due to contamination.[12]

Upstream Fiscal Tools

Denmark has implemented a variety of upstream fiscal tools to develop the circular economy and encourage a life cycle perspective to be taken by economic actors in an attempt to decouple economic growth from resource use and associated environmental impacts.

The Danish Business Authority's Fund for Green Business Development

The Danish Business Authority's Fund for Green Business Development invests in innovative green products and services, the development of new green business models, as well as the establishment of green industrial symbioses. The Fund is comprised of three core programmes as follows.

Fund for Green Business Development

The Fund for Green Business Development has been promoting resource efficiency in Danish businesses by giving grants to selected businesses. The Fund has specifically focused on businesses in the circular and sharing economy. Over the period 2013–2015, the Fund invested over DKK 54 million in 33 projects that were related to six themes:

1. Development of new green business models
2. Product innovation and redesign of products
3. Promotion of sustainable materials in product design
4. Sustainable transition in the textile and fashion industry
5. Reducing food waste
6. Sustainable bio-based products using non-food biomass[13]

New Green Business Models

The Fund for Green Business Development, in partnership with the Danish Regions and the Regional Municipality of Bornholm, has established the New Green Business Models accelerator programme. The aim of the programme is to promote the development of new green business models that generate economic growth for the business and reduce their consumption of resources and environmental impact at the same time. The programme has two phases:

- *Phase one*: Businesses can apply for up to DKK 250,000 to develop a green business model that describes the concept and future cash flows.
- *Phase two*: Businesses can receive support for testing and executing the model in its activities. In this phase, the Danish Regions and the Regional Municipality provide up to DKK 1 million in support with regional support linked to the business' geographical location.[14]

Green Industrial Symbioses

The Green Industrial Symbioses programme ran from 2013–2015 with the aim of promoting competitiveness and resource efficiency through symbioses: a form of commercial collaboration in which one business's residual product is reused as a resource by another business with significant financial and environmental benefits for both parties taking part in the collaboration. To facilitate the development of industrial symbioses, a Task Force for Green Industrial Symbioses was formed to provide free advice to businesses seeking to take part. The Task Force was comprised of consultants with a high level of technical proficiency and commercial expertise. The Task Force provided a range of services to businesses including

- Conducting free resource checks
- Individual matchmaking
- Matchmaking events
- Drafting action plans
- Providing assistance in applying for subsidies[15,16]

Task Force for Resource Efficiency

The Danish Business Authority's Task Force for Resource Efficiency was established in 2014 to increase the competitiveness of the Danish economy by reviewing existing regulations affecting resource productivity and circular economy practices and identifying barriers as well as solutions. The Task Force conducted explorative studies of the experiences and daily work of businesses to understand how barriers manifest themselves and affect businesses' behaviour. The studies also assessed rules and how they were administered, and the assistance businesses received to navigate them. In 2015, the Task Force identified barriers to increasing resource efficiency and over the 2016–2017 period established solution teams for each selected barrier identified to find the most effective way for businesses to use their material inputs and water more efficiently. Using an iterative process, the Task Force alternated between business studies, other analyses, and developed solutions in dialogue with businesses and other relevant authorities.[17]

Danish Green Investment Fund

The Danish Green Investment Fund provides financing options for green projects that contribute towards the sustainable development of Danish society. The companies that implement green investments will experience cost savings, which in turn can enhance their competitiveness. In addition, the activities of the Fund will have positive environmental effects from environmental savings, increased production of renewable energy sources, more resource efficiency, and better waste recycling. The Fund co-finances projects established by companies of all sizes and types in the areas of environmental savings, renewable energy sources, and resource efficiency. The Fund can finance up to 60 percent of the total costs associated with the project with loan amounts ranging from DKK 2 million up to DKK 100 million with a maturity up to 30 years. Each loan application is assessed against the four financing criteria:

1. *Green impact*: The project must contribute to the sustainable development of society and it must comprise a documented sustainable effect, which can be continuously measured and reported.

2. *Economy*: The project must be economically viable. The repayment of the loan should be within the agreed time span. As such, technologies that have not been proven commercially viable will not receive financing.
3. *Market potential/demand*: It should be possible to place the project in a context, where product, knowledge, and/or technology are scalable and therefore contribute to growth; for example, in the form of increased export volume.
4. *Socioeconomic return*: The financing activities of the Fund must contribute to economic growth and job creation.

Environmental Technology Development and Demonstration Programme Fund

The Environmental Technology Development and Demonstration Programme Fund is a DKK 100 million fund that promotes development and application of eco-efficient solutions addressing prioritised environmental challenges while supporting economic growth and employment. Funding is available for

- Developing, testing, and demonstrating new environmental technology in full-scale installations, new buildings, or construction projects
- Establishing innovation partnerships that actively promote cooperation between relevant stakeholders
- Launching projects that support international and bilateral cooperation on activities in environment and innovation
- Building and disseminating knowledge within and about the environmental technology sector[18]

Energy Technology Development and Demonstration Program

The Energy Technology Development and Demonstration Program supports private companies and universities in developing and demonstrating new energy technologies. The programme, with total funding of DKK

170 million, supports renewable energy technologies, energy efficiency technologies, conversion technologies including fuel cells and hydrogen, integration of energy systems including storage, more efficient methods of recovering oil and gas, and storage of CO_2.[19] All project applications are assessed against a criterion summarised in Table 7.1.

Table 7.1 Energy Technology Development and Demonstration Program criteria

Criteria	Emphasis
Project Objective	The project's objective and the state-of-the-art technology is clearly described
Project Time Schedule, Structure and Feasibility	The content of the project is clearly described with clear milestones
Effects of the Project	The project's relevance and potential in relation to the overall programme
Dissemination	It is clear how the project results will be disseminated
Organisation a. Security of supply b. Independence of fossil fuels c. Climate and environmental concerns d. Cost-effectiveness e. Growth and green jobs f. Research prepares development and demonstration of energy technology	The applicant must for the specified efficiency targets qualitatively and with respect to (d) and (e) quantify in what way and to what extent the project is expected to make contributions. Projects do not necessarily have to contribute to all points
Budget and Finance	The project budget is reasonable
Incentive Effect	The project will not be completed if no support is granted
Market Potential	The competitive situation and the expected market for the technology developed is clearly described
Value Proposition	The final target group and the project's added value (economy, comfort, functionality, etc.) is clearly described and the anticipated effects are clear
Research-Technical Assessment	Projects with research content include a clear description
Public-Private Cooperation	The project includes cooperation between companies and public research institutes

Danish Energy Agency. 2017d. Energy Technology Development and Demonstration Programme (EUDP). Call for proposals EUDP 2017. Available: https://ens.dk/sites/ens.dk/files/Forskning_og_udvikling/call_eudp_2017.pdf

Innovation Fund Denmark

The Innovation Fund Denmark was established in 2014 to fund advances in science and technology for the benefit of economic growth and employment in Denmark. In 2017, the Fund has invested DKK 1.25 billion in new initiatives across six research disciplines of (1) bioresources, food, and lifestyle; (2) trade, service, and society; (3) energy, climate, and environment; (4) production, materials, digitisation, and Information and communications technology (ICT); (5) infrastructure, transport, and construction; and (6) biotech, medical, and health. The Fund invests in large-scale projects, growth projects, and talent-related projects, summarised in Table 7.2.

Fertilizer Account

In Denmark, farmers must enter the Register for Fertilizer Account if their annual turnover relating to agricultural activity is more than DKK 50,000 and they meet at least one of the following conditions of

- Having more than 10 livestock units
- Having more than 1.0 livestock unit per hectare
- Receiving more than 25 tonnes of livestock manure

Farmers who enter the Register are required to prepare a fertiliser plan and keep it for five years, calculate the nitrogen quota for the farm, and submit a fertiliser account. The Fertilizer Account contains information about the following:

- *Area sizes and type of crops*: The area size of the farm is the sum of the cultivated, uncultivated, and set-aside areas.
- *The nitrogen standard for the crop*: All crops are given a nitrogen standard.
- *The calculated nitrogen quota for the farm*: The nitrogen quota of the farm is the sum of the nitrogen quota of each field, where the field nitrogen quota is calculated based on the size and the nitrogen standard. The overall nitrogen quota of the farm provides the amount of

Table 7.2 Innovation Fund Denmark projects

Project type	Description	Call type
Large-Scale Projects	These projects involve investments of more than DKK 5 million and comprise projects along the entire value chain from basic research to the market	Thematic Calls in which investments are made for projects falling under one of the Fund's predefined focus areas Open Calls in which investments are not tied to specific themes but are open to all bright ideas Societal Innovation Partnerships involve alliances of 5–10 partners (enterprises, national research institutes, and public authorities) targeting specific societal challenges
Growth Projects (InnoBooster)	Investments of up to DKK 5 million are made to small and medium-sized enterprises (SMEs) with a viable proposition that has high development potential, and which requires venture capital and sparring to nurture their innovation capacity	InnoBooster Investments between DKK 0.5 million and DKK 5 million are available with two annual application deadlines InnoBooster Investments of less than DKK 0.5 million are available with applications received anytime
Talents	Investments are made to support research and entrepreneurial talents with two types of investments made	Industrial Ph.D./Postdoc Investments offer postgraduate research appointments and support the development of research capacities in promising individuals Entrepreneurial Pilot Investments and offer financial support and sparring for recent graduates with innovative entrepreneurial propositions

Innovation Fund Denmark. 2015. Innovation Fund Denmark 2015 Strategy. Available: https://innovationsfonden.dk/sites/defaultfiles/download/2015/02/10/InnovationsfondensstrategiEN.pdf fertiliser (manure and chemical fertiliser) that can be applied on the farm.

- *Number of livestock units and type of livestock*: Animal type and number and the type of housing, feedstuffs, production, and so on are provided so that the amount of nitrogen in the manure produced can be calculated.
- *Use of fertilisers*: Both livestock manure and chemical fertiliser are used.
- *Delivery of chemical fertiliser*: Farmers must report the amount and type of fertiliser supplied.
- *Exchange of fertiliser or manure*: Farmers can exchange fertiliser with other farmers who are in the Register.
- *Manure and fertiliser stock*: Opening and closing stock for the growing season should be calculated annually.[20]

Each year the Danish AgriFish Agency will visit about 1 percent of the farms and an administrative control is run on around 4 percent of the farmers that submit a Fertilizer Account. Farmers who are registered are then allowed to buy chemical fertiliser without paying tax on fertiliser (DKK 4.47 per kg of nitrogen). Meanwhile, farmers with an annual turnover between DKK 20,000 and DKK 50,000 may voluntarily enter the Register.[21]

Upstream Non-Fiscal Tools

Denmark has implemented a variety of upstream non-fiscal tools to develop the circular economy and encourage a life cycle perspective to be taken by economic actors in an attempt to decouple economic growth from resource use and associated environmental impacts.

Environmental Permits for Industry

The largest and potentially most heavily polluting industrial installations in Denmark fall under the Order of Environmental Permitting, which requires all industrial installations that may cause substantial pollution to apply for an environmental permit if their activities are listed under either Annex 1 or Annex 2 of the Order:

- *Annex 1*: Comprises around 2000 activities with a substantial pollution potential
- *Annex 2*: Comprises a list of activities that require an environmental permit but do not have the same pollution potential or complexity as the activities listed in Annex 1, with standard conditions having been drawn up for around 4400 activities

The environmental permit establishes limits for the discharge of substances that could pollute water, soil, and air, as well as limits for odour, noise, and vibrations. The installation must be able to give an account of organisation, prevention, and precautions in the event of an accident. If an operation ceases, requirements are also imposed, for example, on how environmentally harmful substances are to be disposed of.[22]

Chemicals Initiative 2014–2017

In 2013, the Danish Government signed the Chemicals Initiative 2014–2017 agreement with all the political parties to ensure all children and adults can live without fear of becoming ill from chemicals and ensure a healthy environment. A key component of the initiative is to reduce the amount of substances of high concern that often prevent reuse of materials when the products in which they are contained become waste.

New Facility for Substitution of Chemicals

To promote the substitution of chemicals of concern with less hazardous chemicals or new solutions, a new facility has been established to enable dialogue and knowledge-sharing between researchers, authorities, and enterprises on the sustainable use of chemicals in products and processes in Denmark. The new facility will provide SMEs with innovative opportunities to substitute chemicals of concern in products and materials. SMEs will receive advice on how they can progress in phasing out chemicals of concern, with a digital platform developed to enable enterprises and stakeholders to ask questions and advertise projects. In addition, a new substitution portal—SUBSPORT—with up to 500 cases of substitution will be developed to help enterprises work their way towards substitutions. Overall, the new facility aims to collaborate with the business community to

- Substitute chemicals of concern with chemicals or processes of less concern
- Eliminate from products chemicals that are harmful to health and the environment and which comprise a barrier for subsequent recycling
- Optimise chemical processes so that the use of chemicals of concern is minimised and the amount of waste reduced
- Develop tools for the systematic design of processes and products without the use of chemicals of concern

Overall, the facility will contribute towards the development of a circular economy in which materials and assets in products are recirculated so that the consumption of energy, raw materials, as well as chemicals of concern is reduced as much as possible.[23,24]

Industrial Agreement to Ensure Sustainable Biomass (Wood Pellets and Wood Chips)

To reduce levels of CO_2 emissions, the use of sustainable biomass to replace fossil fuels is an important tool; however, it is essential that only biomass components that have a positive effect on the climate are used. To ensure sustainable biomass is used in combined heat and power (CHP) stations, the Danish District Heating Association and the Danish Energy Association are establishing an industry-initiated voluntary framework for sustainable biomass. The Industrial Agreement to Ensure Sustainable Biomass (Wood Pellets and Wood Chips) ensures that the use of solid biomass for energy production is compliant with the framework of sustainability in terms of the environment, health and safety, and climate where CHP producers themselves are responsible for both documenting and satisfying the requirements for sustainability through a third party, with only plants producing more than 20 MW subject to documentation requirements. Biomass sustainability will be documented through annual reporting on compliance with the report either developed or verified by a third party. The annual report will then be made available on the industry members' websites. In addition, the Danish Energy Association and the Danish District Heating Association will provide links to the annual reports on their respective websites. From 2016 onwards, the CHP stations affected have committed themselves to demonstrating on an annual basis the following proportion of wood pellets and wood chips in compliance with the requirements (Table 7.3):

- *2016*: 40 percent
- *2017*: 60 percent
- *2018*: 75 percent
- *2019*: Fully phased in[25]

Downstream Fiscal Tools

Denmark has implemented a variety of downstream fiscal tools to develop the circular economy and encourage a life cycle perspective to be taken by economic actors in an attempt to decouple economic growth from resource use and associated environmental impacts.

Danish Packaging Tax

Denmark's Packaging Tax aims to reduce waste and increase the reuse and recycling rate of packaging. The tax is comprised of a volume-based tax and a weight-based tax.

The Volume-Based Tax

The volume-based tax, a unit duty that varies with materials and volume, aims to create incentives for the collection and refilling of used packaging

Table 7.3 Industrial Agreement to Ensure Sustainable Biomass requirements

Requirement	Description	Actions
Legality	Legality of forest management and utilisation is safeguarded	• Logging from legally designated areas • Payment of all relevant taxes and duties related to the forest sector • Logging complies with applicable legislation governing the environment and forest areas • Logging respects the rights acquired by prescription and the civil rights of indigenous people • Compliance with the trade and customs legislation governing the forest sector
Protection of the Forests' Ecosystems	Forest management must ensure the preservation of the fundamental conditions of the ecosystem	• Assessment of the environment (water, soil, etc.) impacted related to clearance of wood • Impact assessment of the influence of management on ecosystem and biodiversity • Scheme to minimise negative impact on ecosystem and biodiversity including impact from fertiliser, pesticides, and waste disposal
Maintaining Forests' Productivity and Ability to Contribute to the Global Carbon Cycle	Management of forests must ensure the least negative impact on the forest's productivity and carbon sequestration	• Maintaining the forest's ability to produce wood for future generations • Balancing logging and growth rates • Establishing a system for measuring the forest's productivity • Education and training of producers and subcontractors • Refraining from using wood from forests that cannot be replanted/rejuvenated • Refraining from converting land with forest status • Refraining from converting forests with high carbon content

(continued)

Table 7.3 (continued)

Requirement	Description	Actions
Healthy and Well-Functioning Forests	Forest management must ensure healthy and well-functioning forests	• Maintaining or increasing forest health and vitality • Management of natural processes including forest fires, pests, and disease • Protection against illegal logging and mining operations
Protection of Biodiversity, Sensitive Areas, and Areas Worthy of Preservation	Forest management must ensure protection of biodiversity, sensitive areas, and areas worthy of preservation	• Identification of particularly vulnerable areas or areas which are particularly worthy of preservation • Protection of designated areas through forest management with due consideration to sensitive areas and areas worthy of preservation
Social and Work-Related Rights Respected	Forest management must safeguard respect for social and work-related rights	• Identifying, documenting, and respecting original inhabitants with a traditional or legal forest easement • Establishing complaint mechanisms, if not already available, to regulate disagreements relating to the identified forest easements and working environment • Employees shall be entitled to organise themselves and child labour, forced labour, or discrimination are not permitted • The work must be organised and executed in such a way that the employees' health and safety are taken into due consideration

(continued)

Table 7.3 (continued)

Requirement	Description	Actions
CO_2 Emission Limits from Biomass Value Chain	Biomass may only be used where CO_2 emissions from the biomass value chain in question do not exceed the applicable limits resulting from this Agreement at any given time	• BIOGRACE model is chosen as the method of calculation of CO_2 emissions from the biomass value chain • The limits are in place to ensure a significant CO_2 reduction compared to the reference and will take as its starting point the following criteria: Reduction Percentage with reference to the EU's current applicable guidelines, combined with the most ambitious of suitable European standards. The most ambitious standard in Europe right now is from the UK, which aims for a reduction of 70 percent in 2015, 72 percent in 2020, and 75 percent in 2025
Additional Requirements Targeted at Carbon Cycle, Maintenance of Forest Carbon Stock, Indirect Land-Use Change, and Indirect Wooduse Change	To ensure a climate-appropriate carbon balance in addition to the actions in the above seven items and based on the conclusions of the analysis, the industry aims to not use biomass for a variety of circumstances	• Where there regionally exists an actual alternative demand for high-value production (including the production of timber) • Which comes from trees that are grown on fertile soil, which has been unwisely converted from agriculture to forestry • Is to blame for deforestation in the region • That negatively affects the quantity and quality of forest resources in the medium and long term

Danish Energy Association. 2016. The Danish Industry Agreement for Sustainable Biomass. Available: http://www.danishenergyassociation.com/Theme/BiomassForEnergy.aspx

Table 7.4 Volume-based packaging tax rate

Volume (cl)	<10	10–40	41–60	61–110	111–160
Cardboard or Laminate (DKK/unit)	0.15	0.30	0.52	0.97	1.49
Other (glass, plastic, metal, etc.) (DKK/Unit)	0.22	0.52	0.82	1.49	2.38

Green Budget Europe. 2016. The Danish Packaging Tax. Available: http://green-budget.eu/wp-content/uploads/The-Danish-Packaging-tax_volume_FINAL.pdf

by increasing the price of purchasing new packaging, therefore making it more economically viable to collect and refill used packaging. The more the number of times a package is collected and reused the lower will be the tax on the product (Table 7.4).[26]

Weight-Based Tax

The weight-based tax aims to promote the use of more environmentally friendly packaging, in addition to reducing packaging volumes. While the original weight-based tax was abolished in 2014, it still exists on disposable tableware, PVC foil, and plastic and paper carrier bags. The tax is a marginal tax rate of DKK 22 per kilogram with Danish consumers paying between DKK 2 and DKK 3.50 per bag. Bags are covered by the tax if they have a capacity of at least 5 litres.[27]

Beverage Container Deposit-Return Scheme

Since 1984, Denmark has been operating its Beverage Container Deposit-Return Scheme. Until 2002, the scheme only covered refillable beverage containers to limit waste and encourage the reuse of containers. In 2002, the scheme was extended to non-refillable, non-reusable, and disposal beverage containers; for example, metal cans. In 2005, the scheme was further expanded to include 'ready-to-drink' beverages such as alcopops, energy drinks, and cider products. Since 2008, the Scheme also covers mineral water bottles. Deposits apply to both one-way packaging and refillable bottles that contain beer, carbonated soft drinks, energy drinks, mineral water, iced tea, ready-to-drink beverages, and cider products sold in Denmark. It does not cover fresh squash, juice, cocoa, wine, and spirits. Both importers

Table 7.5 One-way disposable packaging refunds

Part	Type	Refund (DKK)
A	Glass bottles and aluminium cans less than 1 litre	1.00
B	Plastic bottles less than 1 litre	1.50
C	All bottles and cans of 1–20 litres	3.00

Dansk Retursystem. 2017. *Deposit amounts in Denmark* [Online]. Available: https://www.danskretursystem.dk/en/all-about-deposits/deposit-amounts/

Table 7.6 Refillable packaging refunds

Material	Size	Refund (DKK)
Glass bottles	Under 0.5 litres	1.00
Glass bottles	Over 0.5 litres	3.00
Plastic bottles	Under 1 litre	1.50
Plastic bottles	Over 1 litre	3.00

Dansk Retursystem. 2017. *Deposit amounts in Denmark* [Online]. Available: https://www.danskretursystem.dk/en/all-about-deposits/deposit-amounts/

and producers must be registered with Dansk Retursystem if they wish to sell drink products that are included in the scheme. Once registered and a fee paid, the companies can affix the deposit logo to their beverage packaging. The take-back is done via reversed vending machines with the amount refunded to customers depending on the type of material used in the bottles and cans, the volume of each bottle or can, and whether the bottle or can will be recycled or reused[28,29] (summarised in Tables 7.5 and 7.6).

Downstream Non-Fiscal Tools

Denmark has implemented a variety of downstream non-fiscal tools to develop the circular economy and encourage a life cycle perspective to be taken by economic actors in an attempt to decouple economic growth from resource use and associated environmental impacts.

Sustainable Procurement

In Denmark, around DKK 300 billion is spent on public procurement each year while private companies account for even more. This type of purchasing power can be used to promote the production and sales of products with lower environmental impacts throughout the entire product chain. The Ministry of Environment and Food has several initiatives to promote green purchasing in Denmark as follows.

Partnership for Green Public Procurement

The Partnership for Green Public Procurement is a collaboration between public organisations who are committed to making extra efforts in reducing their environmental impact from procurement actions and drive the market in a greener direction. The initiative began in 2006 when the Ministry of Environment and Food and the three largest municipalities, Copenhagen, Aarhus, and Odense, entered into the Partnership for Green Public Procurement. Since then other municipalities and regions have joined the partnership. Today the partnership includes 12 municipalities, 2 regions, a Danish water and wastewater company, and the Ministry of Environment and Food. The objective of the partnership is to develop green, cost-effective purchasing which influences the environment both globally and locally with partnership members obliged to follow jointly specified green procurement objectives, have a procurement policy in which environmental concerns play a significant part, and publish the procurement policy on their respective website. In addition, the Partnership has developed a series of binding goals that are summarised in Table 7.7.[30]

The Forum on Sustainable Procurement

In 2010, the Ministry of Environment and Food established the Forum on Sustainable Procurement with the aim of promoting environmentally conscious and sustainable procurement by professional buyers of goods and services in both public and private sectors. The Forum was officially launched in 2011 with it being open to all with an interest in sustainable procurement with membership being free, informal, and personal with members including professional buyers, managers, suppliers, environmentalists, NGOs, government officials, journalists, researchers, consultants, and professionals. The Forum consists of three levels:

Table 7.7 Partnership for Green Public Procurement goals

Sector	Goals	Targets
Food	Seasonal fruit and vegetables	Buy mainly seasonal fruit and vegetables
	Sustainability	Minimum 60 percent fair trade and/or organic coffee, tea, and sugar
	Packaging	PVC-free
	Transportation	The most sustainable options available
Transportation	Cars and light vans	Must meet Center for Green Transport's recommendations on energy efficiency
	Start-stop function	Vehicles must have this functionality
	Energy	Vehicles must be 10 percent better than applicable energy label
	Noise	Vehicles must have a noise level 3 dB below standard requirement
	Fuel	Determination of individual goals for reducing fossil fuels
	Fleet	Individual mapping of fleet
Building and Construction	New construction of buildings	Over 2012–2016, at least 25 percent of new construction had to be classified as low energy
	Renovation of existing buildings	Efforts are made to reduce energy consumption and environmental impact
	Requirements for construction	Each member shall establish requirements and recommendations for new buildings and refurbishing in the areas of energy (electricity, heat, and renewable), water (pure, dirty, and dewatering), waste, indoor (heating and lighting), and transport (means of transport)
Sustainable Wood	Sustainable wood	Wood and wood-based products must meet state regulations for sustainable wood
	Wood product	Wood-based products must meet the criteria of the Nordic Ecolabel

(continued)

Table 7.7 (continued)

Sector	Goals	Targets
Cleaning	Cleaners and sanitary cleaners, laundry and automatic dishwashing, and hand hygiene products	All products purchased must meet the minimum criteria for the Swan, Flower, or corresponding eco-labels including the Nordic Ecolabel
	Wipes	All wipes must be 100 percent recyclable and meet the criteria of the Swan, Flower, or corresponding eco-label
	Disinfectant	Members should avoid/minimise the use of disinfectant cleaners
	Packaging	All products must have clean dosing instructions
	Cleaning service	Daily cleaning must meet the Swan eco-label criteria
Kids Products	Purchasing	All toys purchased for ages 0–6 years should be without phthalates, brominated flame retardants, harmful heavy metals, organic solvents, carcinogenic azo dyes, formaldehyde, and perfume and fragrances
IT	Computers and monitors	All purchased computers and monitors must meet the energy requirements of the Energy Agency's procurement guidelines
Indoor and Outdoor Lighting	Purchasing	For all purchases of lighting the expected energy costs in the lifetime of the lighting must be included in any bid evaluations
Catering Equipment	Equipment purchases	Purchasing decisions need to include environmental and energy requirements for refrigerators and freezers, coffee machines, dishwashing machines, fans, ventilation systems, ovens, packaging, and cooking utensils
Transport of Food and Textiles	Vehicles	Newly purchased cars and vans must meet minimum European standards, vehicles are to be dimensioned with respect to the task, there must be as low consumption of fuel as possible when purchasing heavy vehicles, and guidelines should be developed to restrict idling
Paper and Printed Matter	Paper	All purchased paper for printing and copying is 100 percent recycled paper and/or meets the criteria of the Nordic Ecolabel or the Flower paper products
	Printed matter	All new purchases of printed matter should satisfy the requirements of the Swan or the Flow eco-label

Green Buying. 2017. *Purchasing goals* [Online]. Available: http://www.gronneindkob.dk/indkoebsmaal/

1. A steering committee consisting of invited political organisations
2. A variety of thematic groups
3. Ordinary members

Members of the Forum can work together and propose the establishment of working groups which can hold their own meetings or events, while the steering committee decides which working groups and events can be held within the Forum. The overall objectives of the Forum are to:

- Create visibility and increase awareness of the benefits of sustainable procurement for society, companies, and organisations
- Put sustainable purchasing on the agenda of decision-makers
- Focus on skills development for buyers and suppliers
- Develop a solid and nationwide network of green procurers
- Collect and disseminate existing tools for green procurement[31,32]

The Responsible Procurer

The Responsible Procurer is an Internet web page where procurers can find environmental as well as ethical standards and social and labour clauses ready to copy and paste into tender documents for many product areas including textiles, furniture, electronics, and so forth. The web page also has Total Cost of Ownership tools for selected product areas including computers, monitors, lighting, and fridges/freezers.[33,34]

Energy Labelling of Buildings

In Denmark, energy labelling of buildings is statutory when selling and letting buildings. Energy labelling serves two purposes: First, the label makes visible the energy consumption of the building and therefore functions as informative labelling when the building is sold or let, and secondly, the energy label provides an overview of which energy-related improvements will be cost-effective (the objective, implementation costs, and the savings to be made on electricity and heating bills). The labelling process is carried out by an energy consultant who determines the energy consumption of the building in accordance with standard conditions for weather, household size, operational hours, and habits of consumption. The calculated consumption is a precise indicator for the energy-related quality of a building, in contrast to the actual consumption. The scale of the label spans from A to G and corresponds to other known energy-using products, with G-labelled buildings consuming the most energy. Dwellings, public buildings, and buildings for commercial and service use are all covered by the regulations on energy labelling.[35,36]

Eco-Management and Audit Scheme

The Eco-Management and Audit Scheme (EMAS), developed by the Ministry of Environment and Food, is a voluntary environmental management system for all types of businesses. An EMAS-certified company works continuously to reduce direct environmental impacts in terms of energy and resource consumption, waste production, and emissions (greenhouse gases, air pollution, and wastewater). Furthermore, EMAS-certified companies work to reduce the company's indirect environmental impacts by influencing others, for example, through dialogue with suppliers and customers.[37]

Mandatory Corporate Social Responsibility Reporting

In 2008, the Danish parliament adopted the "Act amending the Danish Financial Statement Act (Accounting for corporate social responsibility (CSR)) in large businesses)", which mandates that large businesses in Denmark account for their work on CSR. In 2013, a new requirement was introduced into the law making it mandatory for businesses to also expressly account for their policies in reducing their climate impact. The businesses that are covered by the statutory requirement must state

- The business's CSR policies, including any standards, guidelines, or principles for CSR used
- How the business translates its CSR policies into action, including any systems or procedures used

- The business's evaluation of what has been achieved through the CSR initiatives during the financial year and any expectations it has regarding future initiatives[38]

Case Study Summary

Denmark has one of the strongest economies in Europe with a balanced budget, stable currency, low-inflation rate, and healthy economic growth. The country's economy is small, yet open to trade with countries inside and outside of Europe. Denmark has a large agricultural sector that produces enough food to feed a population three times its size. Other main industries include engineering, industry, and energy. Nonetheless, Denmark is experiencing a variety of challenges to its linear economy. A quarter of Denmark's population is exposed to particulate matter concentrations that are above EU air quality objectives. The country is at risk of increased heavy precipitation events due to a changing climate as well as an increase in frequency of heatwaves. While there has been a downward trend in consumption of fossil fuels, consumption will soon increase due to a decline in energy improvements, increased demand for electricity from new data centres, and a halt in the installation of new renewable energy capacity due to the end of state subsidies. The country's population is projected to increase significantly over the next half-century with the largest increase expected to be in Copenhagen. Waste levels have increased recently due to an increase in quantities of construction and demolition waste while the amount of municipal waste generated per capita has increased over the past 20-plus years. Finally, Denmark faces water quality challenges with frequent well-closures due to groundwater contamination from multiple sources.

Upstream

Denmark has implemented a variety of upstream fiscal and non-fiscal tools to develop the circular economy and encourage a life cycle perspective to be taken by economic actors in an attempt to decouple economic growth from resource use and associated environmental impacts. Some of the main tools are as follows:

- The Fund for Green Business Development supports the development of innovative eco-efficient products and services as well as new green business models. The initiative also established green industrial symbioses to promote competitiveness and resource efficiency.
- The Danish Task Force for Resource Efficiency was established to increase the competitiveness of the Danish economy. The Task Force reviewed existing regulations that hampered the development of the circular economy and developed solutions to overcome specific barriers to businesses using material inputs and water more efficiently.
- The Danish Green Investment Fund provides financing options for green projects that contribute towards the sustainable development of Danish society. Companies that receive co-funding to implement green investments will experience cost savings as well as positively impact the environment through resource efficiency and better waste recycling.
- The Environmental Technology Development and Demonstration Fund provides funding for the development and application of eco-efficient solutions that address environmental challenges as well as support economic growth and employment.
- The Energy Technology Development and Demonstration Program provides support to companies and universities developing and demonstrating new energy technologies including renewable energy systems and energy efficiency technologies.
- Innovation Fund Denmark funds advances in science and technology, with a focus on areas including bioresources, energy, climate and environment, production, and materials. The fund invests in all types of projects ranging from large-scale projects to entrepreneurial graduate projects.
- Denmark's Fertilizer Account mandates that farmers who meet certain conditions must prepare a fertiliser plan as well as submit a fertiliser account. Farmers who are registered are then allowed to buy chemical fertilisers without paying a tax on fertiliser. Farmers who have a small turnover can voluntarily enter the Register and receive the tax-relief too.
- Industrial installations that impact the environment are required to have an environmental permit that establishes limits for the discharge of substances that could pollute water, soil, and air.

- Denmark's Chemical Initiative has established a new facility that enables dialogue and knowledge-sharing between stakeholders on the sustainable use of chemicals in products and processes and provides SMEs with advice on how they can phase out chemicals of concern. The initiative will also develop a digital platform that enables businesses and stakeholders to ask questions and advertise projects as well as a portal that contains cases of substitutions.
- The Industrial Agreement to Ensure Sustainable Biomass is an industry-initiated voluntary framework that ensures only sustainable biomass that has a positive effect on the climate is used in CHP stations. Biomass sustainability will be documented through annual reporting with reports either developed or verified by a third party with industry partners making the reports available on their respective websites.

Downstream

Denmark has implemented a variety of downstream fiscal and non-fiscal tools to develop the circular economy and encourage a life cycle perspective to be taken by economic actors in an attempt to decouple economic growth from resource use and associated environmental impacts. Some of the main tools are as follows:

- Denmark's Packaging Tax aims to reduce waste and increase the reuse and recycling rate of packaging. The tax is divided into a volume-based tax, which aims to incentivise the collection and refilling of used packaging by increasing the price of purchasing new packaging, and a weight-based tax, which aims to promote the use of more environmentally friendly packaging.
- Denmark operates a deposit-return scheme for beverage containers. The scheme applies to both one-way packaging and refillable bottles sold. The take-back is done via reversed vending machines with the amount refunded to customers depending on the type of material used in the bottles and cans, the volume of each bottle or can, and whether the bottle or can will be recycled or reused.

- The Partnership for Green Public Procurement initiative is a collaboration between public organisations at the national and regional level who are committed to reducing their environmental impact from procurement actions and drive the market in a greener direction. Partnership members are obliged to have in place a procurement policy that considers environmental concerns and make the policy publicly available.
- The Forum on Sustainable Procurement promotes environmentally conscious and sustainable procurement by professional buyers of goods and services in both public and private sectors. The Forum, open to all interested organisations, increases awareness of the benefits of sustainable procurement and collects and disseminates existing tools for green procurement.
- The Responsible Procurer web page has been developed so procurers can find environmental clauses that are ready to be inserted into tender documents for a variety of product areas as well as Total Cost of Ownership tools for a selection of products.
- In Denmark, energy labelling of buildings is mandatory when selling or letting buildings so its energy consumption is visible to buyers and renters and informative in that the label enables an overview of which energy-related improvements can be implemented cost-effectively.
- The EMAS is a voluntary environmental management system that is available for all types of businesses. An EMAS-certified company works continuously to reduce environmental impacts including reducing energy and resource consumption as well as waste. EMAS-certified companies can also influence their suppliers and customers to reduce their environmental impacts.
- All large companies must state their CSR policies including standards, guidelines, or principles for CSR used, how they translate their policies into actions, an evaluation of what has been achieved during the last financial year, and expectations it has regarding future initiatives.

Overall, Denmark has implemented a variety of upstream and downstream fiscal and non-fiscal tools to develop the circular economy and encourage a life cycle perspective to be taken by economic actors in an attempt to decouple economic growth from resource use and associated environmental impacts. These tools are summarised in Table 7.8.

Table 7.8 Denmark case study summary

Tool	Tool type	Tool title	Description	Upstream/Downstream
Fiscal	Environmental Taxes and Charges	Fertilizer Account	Farmers who meet certain conditions must prepare a fertiliser plan and submit a fertiliser account. They can then buy chemical fertilisers without paying a tax on fertiliser	Upstream
		Packaging Tax	Aims to reduce waste and increase the reuse and recycling rate of packaging. The tax is divided into a volume-based tax and a weight-based tax	Downstream
		Beverage Container Deposit-Return Scheme	The scheme applies to both one-way packaging and refillable bottles. The amount refunded to customers depends on the type of materials used, the volume of each bottle or can, and whether the bottle or can will be recycled or reused	Downstream
	Subsidies and Incentives	Fund for Green Business Development	Supports the development of innovative eco-efficient products and services and green business models. The initiative also established green industrial symbioses	Upstream
		Danish Green Investment Fund	Provides financing options for green projects that contribute towards the sustainable development of Danish society	Upstream
		Environmental Technology Development and Demonstration Fund	Provides funding for the development and application of eco-efficient solutions that address environmental challenges	Upstream
		Innovation Fund Denmark	Funds advances in science and technology in areas including bioresources, energy, climate and environmental, production, and materials	Upstream

(continued)

Table 7.8 (continued)

Tool	Tool type	Tool title	Description	Upstream/ Downstream
Non-Fiscal	Regulations	Task Force for Resource Efficiency	Established to review regulations that hampered the development of the circular economy as well as developed solutions that overcame specific barriers	Upstream
		Environmental Permits for Industry	The largest and most heavily polluting industrial installations are required to have an environmental permit	Upstream
		Mandatory CSR Reporting	Mandatory that large companies account for their CSR policies in reducing climate impacts	Downstream
	Green Public Procurement	Partnership for Green Public Procurement	Collaboration between public organisations at the national and regional level to reduce their environmental impact from procurement actions and drive the market in a greener direction	Downstream
	Enhancing Business Competitiveness	Energy Technology Development and Demonstration Program	Provides financial support to companies and universities developing and demonstrating new energy technologies	Upstream
	Raising Industry Awareness and Capacity	Chemicals Initiative 2014-2017	A new facility enables dialogue and knowledge-sharing on the sustainable use of chemicals	Upstream
	Industry-Based Standards	Industrial Agreement to Ensure Sustainable Biomass	An industry-initiated voluntary framework that ensures only sustainable biomass is used in CHP stations	Upstream
	Eco-Labels and Certification	Energy Labelling of Buildings	Energy labelling is mandated when selling or letting buildings so its energy consumption is visible to buyers and renters	Downstream
	Greening the Supply Chain	Forum on Sustainable Procurement	Aims to promote environmentally conscious and sustainable procurement by professional buyers in both public and private sectors	Downstream
		The Responsible Procurer	Internet web page for procurers to find ready-to-use environmental clauses for tender documents as well as Total Cost of Ownership tools	Downstream
	Information-Based Tools	EMAS	A voluntary environmental management system that is available for all types of businesses	Downstream

Notes

1. Ministry of Foreign Affairs of Denmark. 2017. *Danish Business and Economy* [Online]. Available: http://ukraine.um.dk/en/about-denmark/danish-business-and-economy.
2. OECD. 2017. Denmark – Economic forecast summary (June 2017). Available: http://www.oecd.org/economy/denmark-economic-forecast-summary.htm.
3. European Commission. 2017. Economic forecast for Denmark. Available: https://ec.europa.eu/info/business-economy-euro/economic-performance-and-forecasts/economic-performance-country/denmark/economic-forecast-denmark_en.
4. Ministry of Foreign Affairs of Denmark. 2017. *Danish Business and Economy* [Online]. Available: http://ukraine.um.dk/en/about-denmark/danish-business-and-economy.
5. European Environment Agency. 2014. Denmark air pollution country fact sheet Available: https://www.eea.europa.eu/themes/air/air-pollution-country-fact-sheets-2014/denmark-air-pollutant-emissions-country-factsheet/view.
6. Ministry of Environment and Food/Danish Agency for Water and Nature Management. 2017. *Denmark's future climate* [Online]. Available: http://en.klimatilpasning.dk/knowledge/climate/denmarksfutureclimate.aspx.
7. Danish Energy Agency. 2017a. Denmark's energy and climate outlook 2017. Available: https://ens.dk/sites/ens.dk/files/Analyser/denmarks_energy_and_climate_outlook_2017.pdf.
8. Statistics Denmark. 2017. *Population and population projections* [Online]. Available: http://www.dst.dk/en/Statistik/emner/befolkning-og-valg/befolkning-og-befolkningsfremskrivning.
9. City of Copenhagen. 2015. City of Copenhagen municipal plan 2015. The coherent city. Available: https://kp15.kk.dk/sites/kp15.kk.dk/files/municipal_plan_2015.pdf.
10. Ministry of Environment and Food of Denmark. 2016. Waste statistics 2014. Available: http://www2.mst.dk/Udgiv/publications/2016/12/978-87-93529-48-9.pdf.
11. EUROSTAT. 2017. *Municipal waste statistics* [Online]. Available: http://ec.europa.eu/eurostat/statistics-explained/index.php/Municipal_waste_statistics.

12. Ministry of Environment and Food. 2014. Water supply in Denmark. Available: http://www.geus.dk/program-areas/water/denmark/vand forsyning_artikel.pdf.

13. Danish Business Authority. 2017b. *Fund for green business development* [Online]. Available: https://groenomstilling.erhvervsstyrelsen.dk/fund-green-business-development.

14. Danish Business Authority. 2017c. *Green business models* [Online]. Available: https://groenomstilling.erhvervsstyrelsen.dk/green-business-models.

15. Danish Business Authority. 2017d. *Green industrial symbiosis* [Online]. Available: https://groenomstilling.erhvervsstyrelsen.dk/green-industrial-symbiosis.

16. Danish Business Authority. 2017f. *Service offered to businesses* [Online]. Available: https://groenomstilling.erhvervsstyrelsen.dk/offer-businesses.

17. Danish Business Authority. 2017a. *Danish Task Force for Resource Efficiency* [Online]. Available: https://groenomstilling.erhvervsstyrelsen.dk/danish-task-force-resource-efficiency.

18. Danish Environmental Protection Agency. 2017. Consolidated Act on the Environmental Technology Development and Demonstration Programme. Available: http://ecoinnovation.dk/media/182014/bekend-tgoerelse-af-lov-om-mudp_eng_febr2017.pdf.

19. Danish Energy Agency. 2017b. *Energy Technology Development and Demonstration Program* [Online]. Available: https://ens.dk/en/our-responsibilities/research-development/eudp.

20. Anders Nemming and Rune Ventzel Hansen. Fertilizer accounts in Denmark. HELCOM workshop, 2015 Oldenburg. Ministry of Food, Agriculture and Fisheries of Denmark.

21. Ibid.

22. Ministry of Environment and Food of Denmark. 2017d. *Environmental permits for industry* [Online]. Available: http://eng.mst.dk/trade/industry/environmental-permits-for-industry/.

23. Ministry of Environment and Food of Denmark. 2017a. *Chemicals initiatives 2014–2017* [Online]. Available: http://kemikalieindsatsen.dk/english/.

24. Ministry of Environment and Food of Denmark. 2017b. *Circulating resources* [Online]. Available: http://kemikalieindsatsen.dk/english/ressourcer-i-kredslob/.

25. Danish Energy Association. 2016. The Danish Industry Agreement for Sustainable Biomass. Available: http://www.danishenergyassociation. com/Theme/BiomassForEnergy.aspx.
26. Green Budget Europe. 2016. The Danish Packaging Tax. Available: http://green-budget.eu/wp-content/uploads/The-Danish-Packaging-tax_volume_FINAL.pdf.
27. Ibid.
28. PRO EUROPE Packaging Recovery Organisation. 2017. *Denmark* [Online]. Available: http://www.pro-e.org/Denmark.
29. Dansk Retursystem. 2017. *Deposit amounts in Denmark* [Online]. Available: https://www.danskretursystem.dk/en/all-about-deposits/deposit-amounts/.
30. Ministry for Environment and Food of Denmark. 2017. *Partnership for Green Public Procurement* [Online]. Available: http://eng.mst.dk/sustainability/sustainable-consumption-and-production/sustainable-procurement/partnership-for-green-public-procurement/.
31. Ministry of Environment and Food. 2017. *The Forum on Sustainable Procurement* [Online]. Available: http://eng.mst.dk/sustainability/sustainable-consumption-and-production/sustainable-procurement/forum-on-sustainable-procurement/.
32. Forum for Bæredygtige Indkøb. 2017. *Forummets mål* [Online]. Available: http://www.ansvarligeindkob.dk/forummet/forummets-maal/.
33. Ministry of Environment and Food of Denmark. 2017e. *Sustainable procurement* [Online]. Available: http://eng.mst.dk/sustainability/sustainable-consumption-and-production/sustainable-procurement/.
34. Local Government Denmark. 2017. *Totalomkostninger* [Online]. Available: http://csr-indkob.dk/compra/totalomkostninger/.
35. Danish Energy Agency. 2017c. *Energy labels for buildings* [Online]. Available: https://ens.dk/en/our-responsibilities/energy-labels-buildings.
36. Building Rating. 2017. *Denmark's building energy label* [Online]. Available: http://www.buildingrating.org/graphic/denmarks-building-energy-label.
37. Ministry of Environment and Food of Denmark. 2017c. *EMAS and Environmental Management Systems in Denmark* [Online]. Available: http://eng.mst.dk/sustainability/sustainable-consumption-and-production/emas-and-environmental-management/.
38. Danish Business Authority. 2017e. *Legislation* [Online]. Available: http://csrgov.dk/legislation.

References

Anders, Nemming, and Rune Ventzel Hansen. *Fertilizer accounts in Denmark.* HELCOM workshop, 2015 Oldenburg. Ministry of Food, Agriculture and Fisheries of Denmark.

Building Rating. 2017. *Denmark's building energy label* [Online]. Available: http://www.buildingrating.org/graphic/denmarks-building-energy-label.

City of Copenhagen. 2015. City of Copenhagen municipal plan 2015. The coherent city. Available: https://kp15.kk.dk/sites/kp15.kk.dk/files/municipal_plan_2015.pdf.

Danish Business Authority. 2017a. *Danish task force for resource efficiency* [Online]. Available: https://groenomstilling.erhvervsstyrelsen.dk/danish-task-force-resource-efficiency.

———. 2017b. *Fund for green business development* [Online]. Available: https://groenomstilling.erhvervsstyrelsen.dk/fund-green-business-development.

———. 2017c. *Green business models* [Online]. Available: https://groenomstilling.erhvervsstyrelsen.dk/green-business-models.

———. 2017d. *Green industrial symbiosis* [Online]. Available: https://groenomstilling.erhvervsstyrelsen.dk/green-industrial-symbiosis.

———. 2017e. *Legislation* [Online]. Available: http://csrgov.dk/legislation.

———. 2017f. *Service offered to businesses* [Online]. Available: https://groenomstilling.erhvervsstyrelsen.dk/offer-businesses.

Danish Energy Agency. 2017a. Denmark's energy and climate outlook 2017. Available: https://ens.dk/sites/ens.dk/files/Analyser/denmarks_energy_and_climate_outlook_2017.pdf.

———. 2017b. *Dnergy Technology Development and Demonstration Program* [Online]. Available: https://ens.dk/en/our-responsibilities/research-development/eudp.

———. 2017c. *Energy labels for buildings* [Online]. Available: https://ens.dk/en/our-responsibilities/energy-labels-buildings.

———. 2017d. Energy Technology Development and Demonstration Programme (EUDP). Call for proposals EUDP 2017. Available: https://ens.dk/sites/ens.dk/files/Forskning_og_udvikling/call_eudp_2017.pdf.

Danish Energy Association. 2016. The Danish Industry Agreement for Sustainable Biomass. Available: http://www.danishenergyassociation.com/Theme/BiomassForEnergy.aspx.

Danish Environmental Protection Agency. 2017. Consolidated Act on the Environmental Technology Development and Demonstration Programme.

Available: http://ecoinnovation.dk/media/182014/bekendtgoerelse-af-lov-om-mudp_eng_febr2017.pdf.

Dansk Retursystem. 2017. *Deposit amounts in Denmark* [Online]. Available: https://www.danskretursystem.dk/en/all-about-deposits/deposit-amounts/.

European Commission. 2017. Economic forecast for Denmark. Available: https://ec.europa.eu/info/business-economy-euro/economic-performance-and-forecasts/economic-performance-country/denmark/economic-forecast-denmark_en.

European Environment Agency. 2014. Denmark air pollution country fact sheet Available: https://www.eea.europa.eu/themes/air/air-pollution-country-fact-sheets-2014/denmark-air-pollutant-emissions-country-factsheet/view.

Eurostat. 2017. *Municipal waste statistics* [Online]. Available: http://ec.europa.eu/eurostat/statistics-explained/index.php/Municipal_waste_statistics.

Forum for Bæredygtige Indkøb. 2017. *Forummets mål* [Online]. Available: http://www.ansvarligeindkob.dk/forummet/forummets-maal/.

Green Budget Europe. 2016. The Danish Packaging Tax. Available: http://green-budget.eu/wp-content/uploads/The-Danish-Packaging-tax_volume_FINAL.pdf.

Green Buying. 2017. *Purchasing goals* [Online]. Available: http://www.gron-neindkob.dk/indkoebsmaal/.

Innovation Fund Denmark. 2015. Innovation Fund Denmark 2015 Strategy. Available: https://innovationsfonden.dk/sites/default/files/download/2015/02/10/InnovationsfondensstrategiEN.pdf.

Local Government Denmark. 2017. *Totalomkostninger* [Online]. Available: http://csr-indkob.dk/compra/totalomkostninger/.

Ministry for Environment and Food of Denmark. 2017. *Partnership for Green Public Procurement* [Online]. Available: http://eng.mst.dk/sustainability/sustainable-consumption-and-production/sustainable-procurement/partnership-for-green-public-procurement/.

Ministry of Environment and Food. 2014. Water supply in Denmark. Available: http://www.geus.dk/program-areas/water/denmark/vandforsyning_artikel.pdf.

———. 2017. *The Forum on Sustainable Procurement* [Online]. Available: http://eng.mst.dk/sustainability/sustainable-consumption-and-production/sustainable-procurement/forum-on-sustainable-procurement/.

Ministry of Environment and Food/Danish Agency for Water and Nature Management. 2017. *Denmark's future climate* [Online]. Available: http://en.klimatilpasning.dk/knowledge/climate/denmarksfutureclimate.aspx.

Ministry of Environment and Food of Denmark. 2016. Waste statistics 2014. Available: http://www2.mst.dk/Udgiv/publications/2016/12/978-87-93529-48-9.pdf.

———. 2017a. *Chemicals initiatives 2014–2017* [Online]. Available: http://kemikalieindsatsen.dk/english/.

———. 2017b. *Circulating resources* [Online]. Available: http://kemikalieindsatsen.dk/english/ressourcer-i-kredslob/.

———. 2017c. *EMAS and Environmental Management Systems in Denmark* [Online]. Available: http://eng.mst.dk/sustainability/sustainable-consumption-and-production/emas-and-environmental-management/.

———. 2017d. *Environmental permits for industry* [Online]. Available: http://eng.mst.dk/trade/industry/environmental-permits-for-industry/.

———. 2017e. *Sustainable procurement* [Online]. Available: http://eng.mst.dk/sustainability/sustainable-consumption-and-production/sustainable-procurement/.

Ministry of Foreign Affairs of Denmark. 2017. *Danish Business and Economy* [Online]. Available: http://ukraine.um.dk/en/about-denmark/danish-business-and-economy.

OECD. 2017. Denmark – Economic forecast summary (June 2017). Available: http://www.oecd.org/economy/denmark-economic-forecast-summary.htm.

PRO EUROPE Packaging Recovery Organisation. 2017. *Denmark* [Online]. Available: http://www.pro-e.org/Denmark.

Statistics Denmark. 2017. *Population and population projections* [Online]. Available: http://www.dst.dk/en/Statistik/emner/befolkning-og-valg/befolkning-og-befolkningsfremskrivning.

8

Natural Resource Management and the Circular Economy in Germany

Introduction

Germany has a EUR 3.134 trillion economy with the main economic sectors being industry (25.9 percent), public administration, defence, education, human health, and social work activities (18.2 percent), and wholesale and retail trade, transport, accommodation, and food service activities (15.8 percent).[1] Economic growth is projected to be 1.9 percent in 2018 with low unemployment and higher government spending underpinning private consumption.[2,3]

Challenges to the Linear Economy

Germany is experiencing a variety of challenges to its traditional linear economy as described below through an assortment of examples.

© The Author(s) 2018
R. C. Brears, *Natural Resource Management and the Circular Economy*,
Palgrave Studies in Natural Resource Management,
https://doi.org/10.1007/978-3-319-71888-0_8

Air Pollution

In 2016, Germany recorded its lowest measured levels of particulate matter pollution since 2000, with only one exceedance of the European Union (EU) limit value (PM10 daily mean value over 50 micrograms per cubic metre ($\mu g/m^3$) on more than 35 days in a year) throughout the whole country at one measuring station. Nonetheless, the 20 $\mu g/m^3$ mean annual guideline value recommended by the World Health Organization (WHO) was exceeded at 24 percent of all measuring stations. Meanwhile, the air in Germany's cities had excessive levels of nitrogen dioxide pollution in 2016. The threshold value of 40 $\mu g/m^3$ on a yearly average was exceeded at about 57 percent of all measuring stations located near major thoroughfares.[4]

Climate Change

By 2071–2100, climate change is likely to result in average temperatures rising and being seasonally differentiated by about +3.5 °C (summer: +1.5 °C to +5 °C; winter: +2 °C to +4.5 °C) compared to 1961–1990. It is expected that precipitation will decline during the summer months with forecasts suggesting the possibility of up to a 25 percent decrease. At the same time, the proportion of heavy precipitation in the total precipitation will increase. During the winter months, total precipitation is expected to increase by up to 20 percent.[5]

Energy

Over the past ten years, electricity generated from renewable sources has tripled in Germany reaching 195 billion kilowatt hours in 2015, representing 31 percent of the country's gross electricity generation. Of this 31 percent, 20 percent came from solar photovoltaics and wind and the other 11 percent from hydropower, biomass, and waste. Nonetheless, 44 percent of the country's electricity production was generated from coal, 11 percent from other fossil fuels, in addition to 15 percent from nuclear energy.[6]

Population Growth

Germany's population is predicted to decline from a 2002 peak of 82 million to 74.6 million by 2050. The percentage of Germans under 15 is set to fall to 13 percent, among the world's lowest, while the share of the population over 60 is expected to rise from 27 percent to 39 percent. This will result in Germany's working population falling over time. The large immigration influx will only ease some of these pressures; on average, it can take five years before 50 percent of working-age refugees are in jobs.[7]

Water

Germany, with an available water supply of 188 billion m³, is a country rich in water resources. Less than 20 percent of the available water supply is used. The public water supply abstracts around 5.1 billion m³ of water per year, while the mining and industrial sector abstracts about 6.8 billion m³. Thermal power plants have the largest water demand, with around 20.7 billion m³ of water used as cooling water for energy production.[8] Groundwater reserves are the most important source of potable water; however, more than 27 percent of groundwater bodies in Germany have nitrate levels that exceed the limit value of 50 mg/l. If nitrate inputs do not decrease soon, water utilities will have to resort to costly treatment methods to remove the nitrate from raw water, which could raise the price of drinking water by between 55 and 76 cents per cubic metre.[9]

Waste

Germany is the third-largest producer of municipal waste in the EU, generating 618 kg per person in 2014. Nonetheless, Germany was one of the first European countries to introduce landfill limiting policies in the 1990s that included schemes for collecting packaging waste, biowaste, and waste paper separately. As such, recycling of municipal waste has increased from 48 percent in 2001 to 62 percent in 2010 and 66 percent in 2015. In contrast, Germany's material recycling rate is just 47 percent.[10,11,12]

Upstream Fiscal Tools

Germany has implemented a variety of upstream fiscal tools to develop the circular economy and encourage a life cycle perspective to be taken by economic actors in an attempt to decouple economic growth from resource use and associated environmental impacts.

Environmental Protection Innovation Program

The Environmental Protection Innovation Program supports the demonstration of large-scale projects that show for the first time how advanced methods for avoiding or reducing environmental pollution can be used. The programme funds outstanding projects that show other companies in the same sector or across the industry how innovative technology can lead to environmental protection. Demonstration projects that are both ecologically and economically successful provide other companies with an incentive to use environmentally friendly technology or develop their own processes that enhance both the economy and the environment. Projects supported include

- Resource efficiency/material saving
- Energy saving, energy efficiency, and the use of renewable energies
- Wastewater treatment/hydraulic engineering
- Waste prevention, recycling, and disposal of old deposits
- Air pollution (including measures to reduce odours)
- Soil protection
- Reduction of noise and vibrations[13]

Application Process

Applicants can submit their project idea in the form of a project sketch for preliminary evaluation by KfW. If the sketch is pre-approved by both KfW and the Federal Environmental Agency, then a formal project application can be made. If KfW and the Federal Environmental Agency make a positive assessment of the formal application, a recommendation will be made to the Federal Ministry for the Environment, Nature Conservation, Building and Nuclear Safety (Federal Ministry for the Environment). If a positive funding decision is made, the applicant will receive an application form from KfW. There are two types of funding available for applicants with no maximum amount:

1. *Investment grant*: The applicant will receive a direct grant of up to 30 percent of eligible expenditure
2. *Interest subsidy for the reduction of a KfW loan*: The applicant will receive an interest-reduced credit of up to 70 percent of the eligible expenditure[14]

Energy Consulting in SMEs Grant

The Energy Consulting in SMEs Grant programme aims to increase the number of energy consultations carried out in small to medium-sized enterprises (SMEs) and to raise existing energy saving potentials through implementation of energy efficiency measures. For companies with an annual energy cost of more than EUR 10,000 a grant that covers 80 percent of eligible consulting costs is provided, including any implementation consulting (up to a maximum of EUR 8000). For companies with annual energy costs of a maximum of EUR 10,000, the grant amounts to 80 percent of eligible consulting costs, including possible implementation consultations (up to a maximum of EUR 1200).[15]

KfW Waste Heat Program

The KfW Waste Heat Program, funded by the Federal Ministry of Economics and Energy, is a loan programme used to promote investments in the modernisation, extension, or construction of plants that prevent or use waste heat (summarised in Table 8.1). Companies can borrow up to EUR 25 million per project, with financing covering up to 100 percent of the eligible investment costs, with the annual interest rate set at 1 percent.[16]

Table 8.1 KfW Waste Heat Program

Investment focus	Examples of investments
Internal Avoidance and Use of Waste Heat	• Process optimisation • Conversion of production processes to energy-efficient technologies to prevent or use waste heat • General insulation or insulation of systems, piping, and fittings • Return of waste heat to the production process • Preheating of other media • Electricity efficiency measures directly related to waste heat
Using Excessive Waste Heat	• Extraction of waste heat • Connection lines for transferring heat
Electricity Generation from Waste Heat	• Organic Rankine Cycle technology • Natural gas expansion turbines
Waste Heat Concept	• Expenses for implementation support and use of external experts

KFW. 2017. *KfW energy efficiency program – Waste heat* [Online]. Available: https://www.kfw.de/inlandsfoerderung/Unternehmen/Energie-Umwelt/F%C3%B6rderprodukte/EE-Abw%C3%A4rme-%28294%29/#1

Upstream Non-Fiscal Tools

Germany has implemented a variety of upstream non-fiscal tools to develop the circular economy and encourage a life cycle perspective to be taken by economic actors in an attempt to decouple economic growth from resource use and associated environmental impacts.

Germany's Cluster Initiatives

Germany's cluster initiatives have paved the way for innovations in numerous technology and business fields. With the aim of supporting their business activities, many national cluster policy measures have been initiated and implemented since the 1990s. Currently, support is provided by the Federal Ministry of Economics and Technology with its Go-cluster programme. This is on top of the Federal Ministry of Education and Research's Leading-Edge Cluster Competition under the Federal Government's High-Tech Strategy which established thematic priorities in research and innovation including the digital economy and society, sustainable economy and energy, the innovative workplace, healthy living, intelligent mobility, and civil security.[17,18,19]

Go-cluster Programme

The Go-cluster programme provides support to cluster management organisations with the development of their own innovation clusters. Clusters included in the programme are vanguards of innovation and demonstrate how highly competent Germany is in different industries and technological sectors. To become a member of the Go-cluster programme, German innovation clusters submit a proposal which is assessed against a set of quality criteria that must be met. Following a proposal submission, a personal meeting is arranged with the cluster manager. After successful application and admission, the cluster is then required to participate in the benchmarking process of the European Cluster Excellence Initiative (ECEI) and commit themselves to meeting the quality criteria of the Silver Label of ECEI within two years, with the costs of benchmarking and certification covered in full by the Federal Ministry for Economic Affairs and Energy. Members of the Go-cluster programme receive many advantageous benefits including

- Participation and higher visibility in government economic initiatives
- Public presentations of cluster activities and selected success stories on innovation projects in events, newsletters, and websites
- Networking activities with the most efficient clusters from Germany and Europe
- Participation in seminars on topical matters of clusters and management
- Individual coaching on strategy development
- Entitlement to apply for funds[20]

Leading-Edge Cluster Competition

The Leading-Edge Cluster Competition supported the most efficient clusters in creating growth and jobs. Over three competition rounds in

2008, 2010, and 2012, a high-ranking, independent jury selected a total of 15 clusters with each cluster receiving up to EUR 40 million over a period of up to five years. There were no thematic targets; instead, candidates were selected for the best strategies for future markets in their respective sectors, with the following criteria crucial for the strategy:

- Significant financial participation of businesses and private investors
- Planned projects build on strengths and lead to sustainable changes
- The ability to achieve or consolidate an international top position
- Measures for the development and testing of innovative forms for cooperation including professional cluster management
- Cluster-specific training, qualifications, and promotion of new talent[21]

Cleaner Production Germany

Cleaner Production Germany is an online portal for environmental technology transfer. On the portal, over 3000 publications on research findings and best practice examples can be found across a broad range of cleaner production topics including waste management, air pollution control, sustainable mobility, and climate protection with all technical methods and solutions presented having been independently verified. The portal also provides access to technology providers and contacts and presents selected content in high-quality films and 3D animations.[22]

ÖKOBAUDAT: Life Cycle Assessment Construction Sector

The Federal Ministry for the Environment has created ÖKOBAUDAT which is an online database containing quality-checked life cycle assessment data from the construction sector for all relevant building materials. The database is freely accessible as part of the Assessment System for Sustainable Building, which provides scientific and planning-based evaluation processes for office and administration buildings with the goal of describing and evaluating the ecological quality of the buildings.[23]

eLCA Life Cycle Analysis Tool

The online tool ecological life cycle analysis (eLCA) is an easy-to-use life cycle assessment tool for buildings. The main feature of the tool is an editor for assessing building elements including its materials. By providing all construction material data sets in a preconfigured way it enables a uniformed approach for calculating and comparing life cycle assessments of buildings.[24]

German Federal Ecodesign Award

The German Federal Ecodesign Award is aimed at companies of all sizes, in all sectors, as well as individual designers that have developed a product, service, or concept that demonstrates a high level of innovation, from both a design and environmental perspective. Applications are placed under their appropriate category (Table 8.2) with submissions assessed by experts from the German Federal Environmental Agency, the project advisory board, and the jury. Throughout the assessment, consideration is given to the product's/service's/concept's life cycle from the preliminary stages of production to manufacturing, distribution, use, reuse/recycle, and disposal. The repercussions for everyday culture and consumer behaviour are also considered. Winners of each category receive an

Table 8.2 German Federal Ecodesign Award categories

Category	Description
Product	Submissions to the Product category must be available on the German market or exist as market-ready prototypes that will soon be released on the German market
Service	This category invites submissions in the form of services and system solutions that are available on the German market
Concept	The concept category invites submissions in the form of pioneering concepts, studies, and pilot studies that demonstrate a high level of innovation both from a design and environmental perspective
Young Talent	Students and recent graduates are invited to compete in the Young Talent award

Bundespreis. 2017b. *Competition categories* [Online]. Available: https://www.bundespreis-ecodesign.de/en/wettbewerb/2017/kategorien.html

award and the right to advertise themselves with the Ecodesign title. After the official presentation of the award, the winning entries take part in a high-profile publicity campaign in a range of media. They also feature in an online exhibition on the Federal Ecodesign Award's website and take part in a touring exhibition. In addition, winners of the Young Talent award receive a cash prize of EUR 1000.[25]

Check-in Energy Efficiency Model Project

In 2016, the Federal Ministry for Economic Affairs and Energy and the German Energy Agency (Dena) created the Check-in Energy Efficiency Model Project to identify and develop energy-saving potentials in the accommodation sector. Over a four-year period, the programme will work with a select group of hotels on various phases of energy consulting, implementation of measures, and monitoring of energy consumption. Dena invited five hoteliers to participate in the programme where the agency will help the participants improve their guest communication on energy efficiency in addition to investing in small and large-scale measures to optimise their operations, for example, modernising their lighting systems or cooling technology. As part of the programme, Dena commissioned a consultant team with expertise in communication, hotel management, and psychology. The goal of the coaching programme is to develop:

- Communication concepts that turn energetic ideas into positive actions
- New ideas, formats, and measures energy-conscious hotels can use for their communication
- Best practices for other hotels implementing energy efficiency measures

At the start of the programme, Mystery Checks were conducted on each of the participating hotels with the consultants checking in and documenting the status quo in the hotel with photos. The visibility of energy efficiency and sustainability as well as the communication with the guests were the focus. After an overnight stay, the testers revealed themselves and explained their impressions of each hotel. In addition to praise and constructive criticisms, the first ideas for change and the timetable for multi-monthly coaching sessions were provided along with milestones to be achieved at three joint workshops involving all the hotels and their respective consulting teams. Over the next few years, the hotels will work on their concepts and ideas for communicating energy efficiency with their customers, for example, in flyers, on the hotel's website, or on social media channels. The hoteliers will also be provided examples of best practices of other hotels communicating energy efficiency. Overall, the five participants are encouraged to think and exchange ideas on questions such as

- What have we already implemented and what are we planning to do around environmental sustainability/energy efficiency?
- How can the hotel tell a story about the role green themes play?
- How can energy efficiency and sustainability become tangible for the guest?
- Which social media channels can be used and what personnel, time, and financial resources can be invested?[26]

Online Energy Efficiency Portal

Dena has created an online energy efficiency portal for offices, supermarkets, hotels, and administrative buildings to find information about energy savings, promotion programmes, and search for qualified energy consultants and experts. Often implementing energy efficiency measures in non-residential buildings is a challenge as technical equipment and architecture is more complex than in residential buildings and the measures vary depending on the type of industry and use. At the same time, there is a wide range of potential options to save energy and so the portal provides figures and background information on the efficiency potential of non-residential buildings as well as practical examples from different sectors.[27]

Prevention of Food Waste in the Catering Sector Guidelines

The focus of the Prevention of Food Waste in the Catering Sector Guidelines is to help the professional event catering sector, which is behind events such as city festivals, exhibition openings, sporting events, and music festivals, plan and consider all steps involved in catering an event that significantly reduces food wastage and how to make good use of leftovers. The guidelines are aimed at both caterers and their staff and suppliers, as well as the customers of the catered event.[28]

Energy Efficiency: Energy Modernization of Retail Buildings

The Energy Efficiency—Energy Modernization of Retail Buildings programme, run by Dena in partnership with the commercial and real estate sector, will facilitate redevelopment projects that are suitable for imitation by other commercial properties in Germany. The projects aim to provide solutions to widespread obstacles the sector faces in implementing energy efficiency projects such as lack of resources, time pressure, lack of compatibility of the construction measures with current operations, or lack of coordination with the building's owner. The projects are intended to show ways to solve the barriers to increasing energy efficiency in renovations.[29] To apply for participation in the programme, participants must be one of the following, and show total energy savings in their application:

- Trading companies in all industries (food, clothing, furniture, cosmetics, sporting goods, toys, electronics, etc.) and sizes (from global players to individual owner-operated shops)
- Owner of commercial buildings (including mixed-use buildings)
- Joint applications between dealers and owners

The selection of participants in the programme is done by a jury of experts, made up of representatives from politics, commerce, real estate, Dena, and experts from the field. Successful applicants will then undertake the following process:

- An expert analysis of the building site will be conducted, involving on-site data collection, with up to 80 percent of the costs covered by the Federal Office of Economics and Export Control's "Energy Consulting in Small and Medium-sized Businesses" programme.
- Based on an on-site assessment, an expert develops a refurbishment concept for the building which offers various possibilities for improvement and reconstruction and the cost-effectiveness of the measures.
- The owner chooses the appropriate package of measures for the building, following which an implementation plan is created.
- Based on the implementation plan a final report is prepared.
- The modernisation measures are then implemented over a two-year period.[30]

Flagships of Energy-Efficient Waste Heat Use

The Flagships of Energy-Efficient Waste Heat Use project intends to make visible to companies in all industries the significant energy efficiency potential of avoiding and utilising waste heat. Currently, the technical knowledge for preventing waste heat is available but targeted communication strategies that inform decision-makers in various companies of the potential of this resource is lacking. To change this, Dena is carrying out a total of ten flagship projects that will prevent and utilise waste heat. The results of these exemplary projects will act as beacons for other industries and businesses to replicate. As part of the project, Dena will provide the ten companies with free consulting, and help with applying for suitable funding as well as advertising.[31]

Downstream Fiscal Tools

Germany has implemented a variety of downstream fiscal tools to develop the circular economy and encourage a life cycle perspective to be taken by economic actors in an attempt to decouple economic growth from resource use and associated environmental impacts.

Deposit-Scheme for One-Way Drinks Packaging

Since 2005, all non-ecologically advantageous one-way drinks packaging with a filling volume of between 0.1 and 3.0 litres is subject to a deposit of 25 cents. The deposit does not apply to fruit and vegetable juices, milk, wine and spirits, and one-way drinks packaging that is considered ecologically advantageous (beverage cartons, polyethylene bags, and stand-up bags). In shops and supermarkets, reverse vending machines enable consumers to enter one-by-one their drink packaging and receive a refund slip which the consumer takes to the cashier. The cashier then pays the claimed refund to the customer as indicated on the refund slip.[32]

Downstream Non-Fiscal Tools

Germany has implemented a variety of downstream non-fiscal tools to develop the circular economy and encourage a life cycle perspective to be taken by economic actors in an attempt to decouple economic growth from resource use and associated environmental impacts.

Electrical and Electronic Equipment Act

The aim of the Electrical and Electronic Equipment Act (ElektroG) is to protect the environment from harmful substances from electrical and electronic equipment and to conserve natural resources by preventing waste and using resources efficiently. The ElektroG requires producers, as well as importers, exporters, and distributors, to assume responsibilities for the entire life cycle of their products. ElektroG also requires local authorities to set up collection points for waste electrical and electronic equipment (WEEE) that was in use in private households. Producers and so on must collect WEEE from these sites and dispose of it properly. Consumers are obliged to collect their WEEE separately from their household waste and can discard it for free at municipal collection points. Alternatively, consumers can take advantage of ElektroG's take-back system offered by producers and resellers of electrical and electronic equipment. Retail stores with at least 400 m² of sales area for electrical and electronic products are required, upon the sale of a new piece of equipment, to take back a used device of the equivalent type (1:1 take-back obligation) free of charge as well as any WEEE in common household quantities, unless the external dimensions of the waste exceed 25 cm. The latter does not require the customer to buy new electrical or electronic equipment.[33]

The Blue Angel

The Blue Angel eco-label guarantees that a product or service meets high standards regarding its environmental, health, and performance characteristics with each product and service evaluated across its entire life cycle. The criteria for each individual product group must be fulfilled by those products and services awarded with the Blue Angel. To reflect technological advances, the Federal Environmental Agency reviews the Blue Angel's criteria's every three to four years. This means companies need to constantly improve the environmental friendliness of their products or services over time.[34] To date, around 12,000 environmentally friendly products and services from around 1500 companies have been awarded the Blue Angel, making them more environmentally friendly compared to other comparable standard products and services.[35] Overall, the Blue Label

- Rewards high-quality and eco-friendly products
- Helps save money with energy- and water-efficient products
- Sets product-specific standards which are continuously improved
- Helps consumers make informed decisions[36]

Awarding of the Blue Angel

Members of the Environmental Label Jury, which consists of 16 people appointed by the Federal Environment Minister, decide which product groups and service sectors should be awarded the Blue Angel. When a product or service is awarded the Blue Label, the Environmental Label Jury adopts the Basic Award Criteria of the Blue Angel for a limited period. The period of validity is usually three to five years. If a company complies with the criteria found in the relevant Basic Award Criteria, an application for the award of the Blue Angel eco-label for the relevant product or service can be submitted to RAL gGmbH, which is the award body for the environmental label and organises the process for developing the relevant award criteria in independent expert hearings involving all relevant interest groups. Successful product or service applications then enter a contract with RAL gGmbH with the contract stipulating how the label can be used and the period of the label's validity.[37,38]

Use of the Blue Angel

In terms of using the environmental label in advertising, companies must ensure that the environmental label is only used in combination with the product that has been certified with the Blue Angel eco-label. The label is provided in German once the Blue Angel has been awarded and an English version can be made available on request, with the German one mandatory in Germany. The Blue Angel Tick List names the three most important benefits offered by the product awarded with the Blue Angel eco-label. The Tick List makes it clear that the Blue Angel considers more than just environmental benefits. The Tick List can also be used on the products and for advertising purposes.[39]

Blue Angel Prize

Since 2012, the Blue Angel Prize has been awarded to companies that support the cause of the Blue Angel eco-label in an extraordinary way and are particularly successful in environmental protection. The prize is offered by the Environmental Label Jury, the Federal Ministry for the Environment, the Federal Environmental Agency, and RAL gGmbH in collaboration with the German Sustainability Award.[40]

Car Label

The Car Label informs consumers with a colour scale how efficient a vehicle is, enabling consumers at a glance to recognise the CO^2 efficiency class of the new car and the costs expected for fuel and vehicle taxes. The label's classification ranges from A+ (green, very efficient) to G (red, least efficient).[41]

Life Cycle Costing

All federal agencies are required to take life cycle costs into consideration when evaluating bids concerning the procurement of products and services that entail energy consumption. To help conduct these assessments the Federal Ministry for the Environment provides two tools:

1. *UBA Excel tool*: This tool allows for the evaluation of up to five different procurement modalities and factors in all key cost categories such as procurement, operating, and disposal costs
2. *UBA product-specific Excel tool*: This tool enables users to calculate the life cycle costs of computers, multi-functional devices, computer screens, computing centres, flooring, refrigerators, and dishwashers[42]

German RETech Partnership

To increase the international market share of German companies in recycling and disposal technologies, the German Recycling Technologies and Waste Management Partnership (German RETech Partnership) was formed in 2011 by the Initiative Recycling and Efficiency Technology of the Federal Environment Ministry. The German RETech Partnership brings together companies and government institutions from the entire field of waste disposal and recycling to

• Improve the export requirements for companies in the recycling and disposal sector

- Strengthen the competitiveness of German SMEs and promote the marketing of sustainable and innovative recycling and efficiency technology
- Promote the transfer of know-how
- Establish a network of ministries, authorities, scientific institutions, and associations to support the export of German recycling and waste disposal technology as well as the transfer of knowledge

Overall, the German RETech Partnership offers

- A neutral platform for companies and institutions that are interested in exporting German waste disposal and recycling technology
- Joint initiatives in selected target markets for the German waste disposal sector
- A platform for partner search in projects
- Exchange of experience in specialist questions
- Support and organisation of events, meetings, and delegation trips at the national and international level[43]

Case Study Summary

Germany's large economy, dominated by industry, has low unemployment and high government expenditure underpinning private consumption. Nonetheless, Germany is experiencing a variety of challenges to its linear economy. While particulate matter pollution is for the most part below EU levels, values for certain types of particulate matter exceed the guideline values set by the WHO. Meanwhile, German cities have excessive levels of nitrogen dioxide. Climate change is likely to result in a significant decrease in total summer precipitation levels. Nonetheless, there will be an increase in heavy precipitation events during the summer, while total precipitation in winter months will increase. Over half the country's electricity production is generated from coal and other fossil fuels. A challenge moving forward in the century is Germany's declining population with the percentage of young set to be one of the lowest in the world, resulting in Germany's working population falling over time. While Germany is rich in water sources, nearly a third of groundwater reserves is contaminated with nitrates. Finally, Germany is the third-largest producer of municipal waste in the EU and while recycling rates of municipal waste has increased steadily over the past nearly two decades the country's material recycling rate is low.

Upstream

Germany has implemented a variety of upstream fiscal and non-fiscal tools to develop the circular economy and encourage a life cycle perspective to be taken by economic actors in an attempt to decouple economic growth from resource use and associated environmental impacts. Some of the main tools include the following:

- The Environmental Protection Innovation Program supports the demonstration of large-scale projects that show for the first-time advanced methods for avoiding or reducing environmental pollution. The programme has two types of funding available with no maximum amount: an investment grant that covers eligible expenditure or an interest subsidy for the loan.
- The Energy Consulting in SMEs Grant programme aims to increase the number of energy consultations carried out in SMEs and encourage energy savings through the implementation of energy efficiency measures.
- The KfW Waste Heat Program provides low-interest financing for large-scale investments in the modernisation, extension, or construction of plants that prevent or use waste heat. Companies can receive financing to cover all eligible investment costs up to a limit.
- Germany's Go-cluster programme provides support to cluster management organisations across a variety of industries and technological sectors. Members of the programme are required to participate in a benchmarking process with the costs of benchmarking and certification fully funded by the federal government. Overall, participants

receive a variety of benefits including participation and high visibility in government economic initiatives, networking opportunities with efficient clusters from Germany and Europe, individual consultation on strategy development, and the entitlement to apply for funding.

- The Leading-Edge Cluster Competition involved a high-ranking independent jury selecting, over three rounds, a total of 15 clusters to receive funding over a five-year period. With no thematic targets, candidates were selected for the best strategies for future markets in their respective sectors.
- The Cleaner Production Germany online portal encourages environmental technology transfer with the portal containing publications on research findings and best practice examples from a range of cleaner production topics with all technical methods and solutions independently verified.
- The federal government's ÖKOBAUDAT is a free online life cycle assessment database that contains quality-checked data sets from the construction sector for all relevant building materials. Users can use the database to determine a building's ecological quality.
- The eLCA life cycle analysis tool for buildings contains construction material datasets in a preconfigured way, enabling a uniformed approach to life cycle assessments of buildings.
- The German Federal Ecodesign Award is aimed at companies of all sizes, from all sectors, as well as individual designers that have developed a product, service, or concept that demonstrates a high level of innovation, from both a design and environmental perspective. Applications in all categories are assessed on the entire life cycle of the product, service, or concept.
- The Check-in Energy Efficiency Model Project has been implemented to identify and develop energy saving potentials in the accommodation sector. To date, five participants have been selected with each receiving energy coaching, help in implementing energy-saving measures, and monitoring of their energy consumption.
- The online energy efficiency portal has been developed for a variety of commercial and administrative buildings to find information about energy savings and promotion programmes as well as search for quali-

fied energy consultants and experts. The portal also provides figures and background information on the efficiency potential of non-residential buildings as well as practical examples from a variety of sectors.

- The Prevention of Food Waste in the Catering Sector Guidelines aims to help the professional event catering sector plan and consider all steps in reducing food wastage and how to make good use of leftovers.
- Dena's Energy-Efficiency—Energy Modernization of Retail Buildings programme, in partnership with the commercial and real estate sector, will facilitate redevelopment projects that are suitable for imitation by other commercial properties in Germany.
- The Flagships of Energy-Efficient Waste Heat Use project intends to make visible the potential of avoiding and utilising waste heat, motivating companies in all industries to use this resource. Ten flagship projects will act as beacons for industries and businesses to replicate, with each company receiving free consulting, and help with applying for suitable funding as well as advertising.

Downstream

Germany has implemented a variety of downstream fiscal and non-fiscal tools to develop the circular economy and encourage a life cycle perspective to be taken by economic actors in an attempt to decouple economic growth from resource use and associated environmental impacts. Some of the main tools include the following:

- All non-ecologically advantageous one-way drinks packaging of a certain volume, excluding a range of drinks, is subject to a deposit. In shops and supermarkets, reverse vending machines enable consumers to deposit their containers in exchange for a refund slip which can be cashed in at the store.
- ElektroG requires producers, as well as importers, exporters, and distributors, to assume responsibilities for the entire life cycle of their

product. The Act requires local authorities to establish collection points for WEEE, with producers and so forth required to collect WEEE from these sites and dispose of it properly. In addition, retail stores of a certain size and above are required, upon sale of a new product, to take back a used device of the equivalent type free of charge, as well as any waste electrical or electronic equipment in common household quantities.

- The Blue Angel eco-label scheme guarantees that labelled products or services meet high standards regarding their environmental, health, and performance characteristics, with each product or service evaluated across its entire life cycle. With technological advances, the criteria used for assessment is reviewed every several years to ensure companies constantly improve the environmental friendliness of their products and services.
- The Car Label informs consumers with a colour scale how efficient a vehicle is, enabling consumers to recognise the carbon dioxide efficiency class of new cars and their fuel costs and vehicle taxes.
- To help federal agencies consider life cycle costs when evaluating procurement bids of products and services entailing energy consumption, the Federal Ministry for the Environment has developed the UBA Excel tool which calculates procurement, operating, and disposal costs and the UBA product-specific Excel tool that enables users to calculate the life cycle costs of a range of specific products.
- To increase the market share of German recycling technologies, the German RETech Partnership was developed to bring together companies and government institutions across the whole sector to strengthen the competitiveness of German SMEs and promote the marketing of innovative waste disposal and recycling technology.

Overall, Germany has implemented a variety of upstream and downstream fiscal and non-fiscal tools to develop the circular economy and encourage a life cycle perspective to be taken by economic actors in an attempt to decouple economic growth from resource use and associated environmental impacts. These tools are summarised in Table 8.3.

Table 8.3 Germany case study summary

Tool	Tool type	Tool title	Description	Upstream/ Downstream
Fiscal	Environmental Taxes and Charges	Deposit-Scheme for One-way Drinks Packaging	All non-ecologically advantageous one-way drinks packaging of a certain volume, excluding a range of drinks, is subject to a deposit	Downstream
	Subsidies and Incentives	Environmental Protection Innovation Program	Funding of large-scale projects that show for the first-time advanced methods for avoiding or reducing environmental pollution	Upstream
		Energy Consulting in SMEs Grant	Grants for energy consultations carried out in SMEs including implementation of energy efficiency measures	Upstream
		KfW Waste Heat Program	Low-interest financing for large-scale investments that prevent or use waste heat	Upstream
Non-Fiscal	Enhancing Business Competitiveness	Cleaner Production Germany	The online portal encourages environmental technology transfer	Upstream
		Energy Efficiency—Energy Modernization of Retail Buildings programme	Dena, in partnership with commercial and real estate partners, will facilitate redevelopment projects that are suitable for imitation by other commercial properties in Germany	Upstream
		Flagships of Energy-Efficient Waste Heat Use	Ten flagship projects will act as beacons for industries and businesses to replicate, with each company receiving free consulting, and help with applying for suitable funding as well as advertising	Upstream
	Cluster Policies	Go-cluster programme	Provides support to cluster management organisations	Upstream
		Leading-Edge Cluster Competition	Clusters were awarded funding to develop the best strategies for future markets in their respective sectors	Upstream
	Raising Industry Awareness and Capacity	Check-in Energy Efficiency Model Project	Helps identify and develop energy saving potentials in the accommodation sector	Upstream
		Prevention of Food Waste in the Catering Sector Guidelines	Helps catering companies plan and consider all steps in reducing food wastage	Upstream

(continued)

Table 8.3 (continued)

Tool	Tool type	Tool title	Description	Upstream/Downstream
	Eco-Labels and Certification	Blue Angel	All labelled products or services meet high standards regarding their environmental, health, and performance characteristics with each product or service evaluated across its entire life cycle	Downstream
		Car Label	Informs consumers on how efficient a vehicle is, enabling consumers to check carbon dioxide efficiency, their fuel costs, and vehicle taxes	Downstream
	Supporting Life-Cycle Analysis	ÖKOBAUDAT	Free online life cycle assessment database that contains quality-checked data sets from the construction sector	Upstream
		eLCA	Enables a uniformed approach to life cycle assessments of buildings	Upstream
		UBA Excel tool	Helps federal agencies calculate procurement, operating, and disposal costs of products and services entailing energy consumption	Downstream
		UBA product-specific Excel tool	Enables federal agencies to calculate the life cycle costs of a range of specific products entailing energy consumption	Downstream
	Environmental Recognition Awards	German Federal Ecodesign Award	Awards for companies and individual designers that have developed a product, service, or concept that demonstrates innovation, from both a design and environmental perspective	Upstream
	Extended Producer Responsibility	ElektroG	Producers, as well as importers, exporters, and distributors assume responsibilities for the entire life cycle of their product	Downstream
	Knowledge Transfer Networks	German RETech Partnership	Brings together companies and government institutions to strengthen and promote the marketing of innovative waste disposal and recycling technology	Downstream
	Information-Based Tools	Online energy efficiency portal	A variety of commercial and administrative buildings can find information about energy savings, promotion programmes, and search for qualified energy consultants and experts	Upstream

Notes

1. European Union. 2017. *Germany* [Online]. Available: https://europa.eu/european-union/about-eu/countries/member-countries/germany_en.
2. European Commission. 2017. European economic forecast. Available: https://ec.europa.eu/info/sites/info/files/ip053_en.pdf.
3. OECD. 2017. Germany – Economic forecast summary (June 2017). Available: http://www.oecd.org/economy/germany-economic-forecast-summary.htm.
4. UBA. 2017a. *Air quality 2016: Nitrogen dioxide still the top pollutant. Lower levels of particulate and ozone pollution* [Online]. Available: https://www.umweltbundesamt.de/en/press/pressinformation/air-quality-2016-nitrogen-dioxide-still-the-top.
5. UBA. 2013. *Expected climate changes* [Online]. Available: https://www.umweltbundesamt.de/en/topics/climate-energy/climate-change-adaptation/impacts-of-climate-change/climate-models-scenarios/expected-climate-changes#textpart-1.
6. EIA. 2016. *Germany's renewables electricity generation grows in 2015, but coal still dominant* [Online]. Available: https://www.eia.gov/todayinenergy/detail.php?id=26372.
7. FT. 2016. *Germany's demographic dilemma* [Online]. Available: https://www.ft.com/content/a4d8316e-8566-11e6-8897-2359a58ac7a5.
8. UBA. 2014. Water management in Germany. Water supply – waste water disposal. Available: https://www.umweltbundesamt.de/sites/default/files/medien/378/publikationen/wawiflyer_uba_en_web.pdf.
9. UBA. 2017f. *Too much fertiliser: drinking water could become more expensive. Price increase of up to 45 per cent expected* [Online]. Available: https://www.umweltbundesamt.de/en/press/pressinformation/too-much-fertiliser-drinking-water-could-become.
10. Eurostat. 2015. Each person in the EU generated 481 kg of municipal waste in 2013. Available: http://ec.europa.eu/eurostat/documents/2995521/6757479/8-26032015-AP-EN.pdf/a2982b86-9d56-401c-8443-ec5b08e543cc.
11. Eurostat. 2017. *Recycling rate of municipal waste* [Online]. Available: http://ec.europa.eu/eurostat/tgm/table.do?tab=table&plugin=1&language=en&pcode=t2020_rt120.
12. Northern Ireland Assembly 2017. Recycling in Germany.

13. UBA. 2017c. *The environmental innovation program* [Online]. Available: http://www.umweltinnovationsprogramm.de/das-umweltinnovations programm.
14. UBA. 2017d. *How is funding supported?* [Online]. Available: http://www.umweltinnovationsprogramm.de/das-uip/wie-wird-gefoerdert.
15. BAFA. 2017. *Energy consulting in SMEs* [Online]. Available: http://www.bafa.de/DE/Energie/Energieberatung/Energieberatung_Mittelstand/energieberatung_mittelstand_node.html.
16. KFW. 2017. *KfW energy efficiency program – Waste heat* [Online]. Available: https://www.kfw.de/inlandsfoerderung/Unternehmen/Energie-Umwelt/F%C3%B6rderprodukte/EE-Abw%C3%A4rme-%28294%29/#1.
17. Clusterportal-BW. 2017. *Cluster policy in Germany* [Online]. Available: https://www.clusterportal-bw.de/en/cluster-policy/cluster-policy-in-germany/.
18. Federal Ministry of Education and Research. 2017a. *The new high-tech strategy* [Online]. Available: https://www.hightech-strategie.de/de/The-new-High-Tech-Strategy-390.php.
19. Federal Ministry of Education and Research. 2014. The new high-tech strategy. Innovations for Germany. Available: https://www.bmbf.de/pub/HTS_Broschuere_eng.pdf.
20. Federal Ministry for Economic Affairs and Energy. 2017. *The "go-cluster" programmes* [Online]. Available: http://www.clusterplattform.de/CLUSTER/Redaktion/EN/Standardartikel/go-cluster.html.
21. Federal Ministry of Education and Research. 2017b. *The top cluster competition* [Online]. Available: https://www.bmbf.de/de/der-spitzencluster-wettbewerb-537.html.
22. Cleaner Production Germany. 2017. *Cleaner production Germany (CPG) – The portal to environmental technology transfer!* [Online]. Available: http://www.cleaner-production.de/index.php/en/.
23. BMUB. 2017a. *Assessment system for sustainable building* [Online]. Available: http://www.nachhaltigesbauen.de/sustainable-building-english-speaking-information/assessment-system-for-sustainable-building.html.
24. BMUB. 2016. Life cycle assessment construction sector. Available: https://www.bmub.bund.de/fileadmin/Daten_BMU/Pools/Broschueren/oekobilanzierung_bauwesen_faltblatt_en_bf.pdf.
25. Bundespreis. 2017a. *The award* [Online]. Available: https://www.bundespreis-ecodesign.de/en/wettbewerb.html.

26. DENA. 2017c. *Model project "Check-in energy efficiency"* [Online]. Available: http://www.effizienzgebaeude.dena.de/best-practice/modellvorhaben-hotels-und-herbergen/das-modellvorhaben/ueber-das-modellvorhaben/.
27. DENA. 2017d. *New online portal provides information on energy efficiency in public and commercial real estate* [Online]. Available: https://www.dena.de/newsroom/meldungen/2017/neues-onlineportal-informiert-zu-energieeffizienz-in-oeffentlichen-und-gewerblichen-immobilien/.
28. UBA. 2016. *Food waste in the catering sector* [Online]. Available: https://www.umweltbundesamt.de/en/topics/waste-resources/waste-management/waste-prevention/food-waste-in-the-catering-sector.
29. DENA. 2017a. *Energetic modernization of retail buildings* [Online]. Available: http://effizienzgebaeude.dena.de/best-practice/modellvorhaben-handelsgebaeude/.
30. DENA. 2017f. *Process and content* [Online]. Available: http://effizienzgebaeude.dena.de/best-practice/modellvorhaben-handelsgebaeude/ablauf-und-inhalte/.
31. DENA. 2017b. *Flagships for waste heat* [Online]. Available: https://www.dena.de/en/topics-projects/projects/energy-systems/flagships-for-waste-heat/.
32. BMUB. 2017b. *Packaging waste* [Online]. Available: http://www.bmub.bund.de/en/topics/water-waste-soil/waste-management/types-of-waste-waste-flows/packaging-waste/.
33. UBA. 2017b. *Electrical and Electronic Equipment Act* [Online]. Available: https://www.umweltbundesamt.de/en/topics/waste-resources/product-stewardship-waste-management/electrical-electronic-waste/electrical-electronic-equipment-act.
34. The Blue Angel. 2017e. *What is behind it?* [Online]. Available: https://www.blauer-engel.de/en/blue-angel/what-is-behind-it.
35. UBA. 2015. *Blue Angel Prize 2015 distinguishes companies serving as role models of environmental and climate protection* [Online]. Available: https://www.umweltbundesamt.de/en/press/pressinformation/blue-angel-prize-2015-distinguishes-companies.
36. The Blue Angel. 2017c. *Our label for the environment* [Online]. Available: https://www.blauer-engel.de/en/our-label-environment.
37. The Blue Angel. 2017a. *Basic award criteria* [Online]. Available: https://www.blauer-engel.de/en/companies/basic-award-criteria.
38. The Blue Angel. 2016. General Information on the life of basic criteria for award of the Blue Angel eco-label Available: https://www.blauer-engel.de/sites/default/files/pages/downloads/survey-all-basic-award-criteria/life-basic-criteria-award.pdf.

39. The Blue Angel. 2017d. *Use of the logo* [Online]. Available: https://www. blauer-engel.de/en/companies/how-do-you-use-blue-angel/use-logo.
40. The Blue Angel. 2017b. *How to use the Blue Angel?* [Online]. Available: https://www.blauer-engel.de/en/companies/how-do-you-use-blue-angel/blue-angel-award.
41. DENA. 2017e. *The Pkw-label* [Online]. Available: https://www.pkw-label.de/pkw-label/das-pkw-label/.
42. UBA. 2017e. *Life cycle costing* [Online]. Available: https://www.umwelt-bundesamt.de/en/topics/economics-consumption/green-procurement/life-cycle-costing.
43. German REtech Partnership. 2017. *About RETech* [Online]. Available: http://www.retech-germany.net/ueber-retech/.

References

BAFA. 2017. *Energy consulting in SMEs* [Online]. Available: http://www.bafa.de/DE/Energie/Energieberatung/Energieberatung_Mittelstand/energieberatung_mittelstand_node.html.
BMUB. 2016. Life cycle assessment construction sector. Available: https://www.bmub.bund.de/fileadmin/Daten_BMU/Pools/Broschueren/oekobilanzierung_bauwesen_faltblatt_en_bf.pdf.
———. 2017a. *Assessment system for sustainable building* [Online]. Available: http://www.nachhaltigesbauen.de/sustainable-building-english-speaking-information/assessment-system-for-sustainable-building.html.
———. 2017b. *Packaging waste* [Online]. Available: http://www.bmub.bund.de/en/topics/water-waste-soil/waste-management/types-of-waste-waste-flows/packaging-waste/.
Bundespreis. 2017a. *The award* [Online]. Available: https://www.bundespreis-ecodesign.de/en/wettbewerb.html.
———. 2017b. *Competition categories* [Online]. Available: https://www.bundespreis-ecodesign.de/en/wettbewerb/2017/kategorien.html.
Cleaner Production Germany. 2017. *Cleaner production Germany (CPG) – The portal to environmental technology transfer!* [Online]. Available: http://www.cleaner-production.de/index.php/en/.
Clusterportal-BW. 2017. *Cluster policy in Germany* [Online]. Available: https://www.clusterportal-bw.de/en/cluster-policy/cluster-policy-in-germany/.

DENA. 2017a. *Energetic modernization of retail buildings* [Online]. Available: http://effizienzgebaeude.dena.de/best-practice/modellvorhaben-handelsgebaeude/.
———. 2017b. *Flagships for waste heat* [Online]. Available: https://www.dena.de/en/topics-projects/projects/energy-systems/flagships-for-waste-heat/.
———. 2017c. *Model project "Check-in energy efficiency"* [Online]. Available: http://www.effizienzgebaeude.dena.de/best-practice/modellvorhaben-hotels-und-herbergen/das-modellvorhaben/ueber-das-modellvorhaben/.
———. 2017d. *New online portal provides information on energy efficiency in public and commercial real estate* [Online]. Available: https://www.dena.de/newsroom/meldungen/2017/neues-onlineportal-informiert-zu-energieeffizienz-in-oeffentlichen-und-gewerblichen-immobilien/.
———. 2017e. *The Pkw-label* [Online]. Available: https://www.pkw-label.de/pkw-label/das-pkw-label/.
———. 2017f. *Process and content* [Online]. Available: http://effizienzgebaeude.dena.de/best-practice/modellvorhaben-handelsgebaeude/ablauf-und-inhalte/.
EIA. 2016. *Germany's renewables electricity generation grows in 2015, but coal still dominant* [Online]. Available: https://www.eia.gov/todayinenergy/detail.php?id=26372.
European Commission. 2017. European economic forecast. Available: https://ec.europa.eu/info/sites/info/files/ip053_en.pdf.
European Union. 2017. *Germany* [Online]. Available: https://europa.eu/european-union/about-eu/countries/member-countries/germany_en.
Eurostat. 2015. Each person in the EU generated 481 kg of municipal waste in 2013. Available: http://ec.europa.eu/eurostat/documents/2995521/6757479/8-26032015-AP-EN.pdf/a2982b86-9d56-401c-8443-ec5b08e543cc.
———. 2017. *Recycling rate of municipal waste* [Online]. Available: http://ec.europa.eu/eurostat/tgm/table.do?tab=table&plugin=1&language=en&pcode=t2020_rt120.
Federal Ministry for Economic Affairs and Energy. 2017. *The "go-cluster" programmes* [Online]. Available: http://www.clusterplattform.de/CLUSTER/Redaktion/EN/Standardartikel/go-cluster.html.
Federal Ministry of Education and Research. 2014. The new high-tech strategy. Innovations for Germany. Available: https://www.bmbf.de/pub/HTS_Broschuere_eng.pdf.
———. 2017a. *The new high-tech strategy* [Online]. Available: https://www.hightech-strategie.de/de/The-new-High-Tech-Strategy-390.php.
———. 2017b. *The top cluster competition* [Online]. Available: https://www.bmbf.de/de/der-spitzencluster-wettbewerb-537.html.

FT. 2016. *Germany's demographic dilemma* [Online]. Available: https://www.ft.com/content/a4d8316e-8566-11e6-8897-2359a58ac7a5.

German Retech Partnership. 2017. *About RETech* [Online]. Available: http://www.retech-germany.net/ueber-retech/.

KFW. 2017. *KfW energy efficiency program – Waste heat* [Online]. Available: https://www.kfw.de/inlandsfoerderung/Unternehmen/Energie-Umwelt/F%C3%B6rderprodukte/EE-Abw%C3%A4rme-%28294%29/#1.

Northern Ireland Assembly. 2017. Recycling in Germany.

OECD. 2017. Germany – Economic forecast summary (June 2017). Available: http://www.oecd.org/economy/germany-economic-forecast-summary.htm.

The Blue Angel. 2016. General Information on the life of basic criteria for award of the Blue Angel eco-label Available: https://www.blauer-engel.de/sites/default/files/pages/downloads/survey-all-basic-award-criteria/life-basic-criteria-award.pdf.

———. 2017a. *Basic award criteria* [Online]. Available: https://www.blauer-engel.de/en/companies/basic-award-criteria.

———. 2017b. *How to use the Blue Angel?* [Online]. Available: https://www.blauer-engel.de/en/companies/how-do-you-use-blue-angel/blue-angel-award.

———. 2017c. *Our label for the environment* [Online]. Available: https://www.blauer-engel.de/en/our-label-environment.

———. 2017d. *Use of the logo* [Online]. Available: https://www.blauer-engel.de/en/companies/how-do-you-use-blue-angel/use-logo.

———. 2017e. *What is behind it?* [Online]. Available: https://www.blauer-engel.de/en/blue-angel/what-is-behind-it.

UBA. 2013. *Expected climate changes* [Online]. Available: https://www.umweltbundesamt.de/en/topics/climate-energy/climate-change-adaptation/impacts-of-climate-change/climate-models-scenarios/expected-climate-changes#textpart-1.

———. 2014. Water management in Germany. Water supply – waste water disposal Available: https://www.umweltbundesamt.de/sites/default/files/medien/378/publikationen/wawiflyer_uba_en_web.pdf.

———. 2015. *Blue Angel Prize 2015 distinguishes companies serving as role models of environmental and climate protection* [Online]. Available: https://www.umweltbundesamt.de/en/press/pressinformation/blue-angel-prize-2015-distinguishes-companies.

———. 2016. *Food waste in the catering sector* [Online]. Available: https://www.umweltbundesamt.de/en/topics/waste-resources/waste-management/waste-prevention/food-waste-in-the-catering-sector.

———. 2017a. *Air quality 2016: Nitrogen dioxide still the top pollutant. Lower levels of particulate and ozone pollution* [Online]. Available: https://www.umweltbundesamt.de/en/press/pressinformation/air-quality-2016-nitrogen-dioxide-still-the-top.

———. 2017b. *Electrical and Electronic Equipment Act* [Online]. Available: https://www.umweltbundesamt.de/en/topics/waste-resources/product-stewardship-waste-management/electrical-electronic-waste/electrical-electronic-equipment-act.

———. 2017c. *The environmental innovation program* [Online]. Available: http://www.umweltinnovationsprogramm.de/das-umweltinnovationsprogramm.

———. 2017d. *How is funding supported?* [Online]. Available: http://www.umweltinnovationsprogramm.de/das-uip/wie-wird-gefoerdert.

———. 2017e. *Life cycle costing* [Online]. Available: https://www.umweltbundesamt.de/en/topics/economics-consumption/green-procurement/life-cycle-costing.

———. 2017f. *Too much fertiliser: Drinking water could become more expensive. Price increase of up to 45 per cent expected* [Online]. Available: https://www.umweltbundesamt.de/en/press/pressinformation/too-much-fertiliser-drinking-water-could-become.

9

Natural Resource Management and the Circular Economy in the Netherlands

Introduction

The Netherlands is the sixth-largest economy in the European Union (EU) and is an important European transportation hub. Some of the largest sectors of the Dutch economy include wholesale and retail trade, transport, accommodation, and food services, which combined is 21 percent of GDP while industry makes up 15.4 percent.[1,2] Around 33 percent of the Netherlands' income is derived from the export of goods and services. One of its most productive export sectors is agriculture which employs only 2 percent of the labour force but provides large surpluses for food processing, making the Netherlands the second-largest agricultural exporter in the world. The economy is projected to show robust growth of just over 2 percent in 2017–2018, but after 2018 a mild slowdown is expected.[3]

© The Author(s) 2018
R. C. Brears, *Natural Resource Management and the Circular Economy*,
Palgrave Studies in Natural Resource Management,
https://doi.org/10.1007/978-3-319-71888-0_9

Challenges to the Linear Economy

The Netherlands is experiencing a variety of challenges to its traditional linear economy as described below through an assortment of examples.

Air Pollution

The emission of several air pollutants has decreased significantly over the period 1990–2014, with reductions of sulphur oxides (−85 percent), nitrogen oxides (−61 percent), and volatile organic compounds (−71 percent). However, air pollution is still a major concern with the European Environment Agency estimating that in 2013 there were around 11,530 premature deaths attributable to fine particulate matter concentrations and an additional 270 to ozone concentration, and over 1820 to nitrogen oxide.[4]

Climate Change

In the winter months, climate change is likely to result in precipitation levels increasing in general and in intensity while during the summer the Netherlands is likely to experience higher intensity of extreme rainfall events, with hail and thunder becoming more severe. In the medium-term for 2030, the temperature is likely to be +1.0 °C higher and mean precipitation +5 percent higher above the reference point of 1981–2010. Increased droughts will lead to water shortages, water quality issues, and salinisation. With the Netherlands being a low-lying country, with some parts below sea level, infrastructure and people will be vulnerable to sea level rise of up to 40 cm by 2050 relative to 1981–2010.[5]

Energy

In the Netherlands, energy consumption is highest for the built environment (33 percent), followed by industry (28 percent), and traffic and transportation (23 percent). The power mix is dominated by natural gas followed by coal and wind, with the country being the second-largest producer and exporter of natural gas in Europe. The share of renewable energy is 5.8 percent, and therefore, it does not appear that the Netherlands will meet its target of having 14 percent of its electricity coming from renewable sources by 2020.[6,7]

Population Growth

In 2017, the Netherlands' population reached 17.1 million, up from 17 million in 2016. Migration and an increase in the number of babies born accounted for around 88,000 and 22,000 of this population growth respectively.[8] The country's four largest cities increased in size too. Amsterdam was the fastest-growing city with its population increasing by over 15,000 to reach a total population of almost 850,000 inhabitants. Rotterdam's population increased by 8000, while the population in both The Hague and Utrecht increased by 4000 each. In contrast, one in five municipalities saw their populations decrease.[9]

Waste

The Netherlands' municipal waste in 2014 was almost the same as in 2013, breaking the downward trend seen in prior years. Overall, the country's municipal waste per capita (527 kg/year/inhabitant) is above the EU average (475 kg/year/inhabitant). The recycling rate, including composting, of municipal waste, is 51 percent, incineration (energy recovery) is 48 percent, and landfilling 1 percent.[10]

Water

The main pressure on the Netherlands' water resources is diffuse pollution that affects 90 percent of all water bodies. The next largest impact on water resources is river management decisions affecting 73 percent of the country's waterways followed by flow regulation and morphological alterations that affect 58 percent and 69 percent of water bodies respectively. Point sources of pollution and abstractions also place significant pressures on water resources in the country, affecting 30 percent and 17 percent of water bodies respectively.[11]

Upstream Fiscal Tools

The Netherlands has implemented a variety of upstream fiscal tools to develop the circular economy and encourage a life cycle perspective to be taken by economic actors in an attempt to decouple economic growth from resource use and associated environmental impacts.

The Green Funds Scheme

The Green Funds Scheme is a tax incentive to encourage individual investors to invest in green projects that benefit the environment such as renewable energy, sustainable construction, and environmental technology. Individuals who invest in a green fund or save money with a financial institution that practices 'green banking' receive a lower interest rate than that of the market rate; however, this is compensated for by a tax exemption.[12]

Tax Exemptions for Green Investments

Green investments are eligible for tax exemptions if the value of the investment does not exceed a certain amount. If the value of the investment exceeds the amount, then the investor will be liable for taxation on the difference. For example, using the green investment tax exemption amount in Table 9.1, if the value of an investment made by an investor without a tax partner on 1 January 2017 is EUR 20,000 then the tax exemption is EUR 20,000, but if the value of the investment is EUR 65,000 then the investor is liable for taxation on the difference between EUR 65,000 and EUR 57,385.[13]

Table 9.1 Green investment tax exemption amount

Year	Without tax partner (EUR)	With tax partner (EUR)
2017	57,385	114,770
2016	57,213	111,426
2015	56,420	112,840
2014	56,420	112,840

Tax and Customs Administration. 2017. *Green investments* [Online]. Available: https://www.belastingdienst.nl/wps/wcm/connect/bldcontentnl/belastingdienst/prive/vermogen_en_aanmerkelijk_belang/vermogen/beleggen_met_belastingvoordeel/groene_beleggingen#

Energy Investment Allowance Tax Incentive

Companies can use the Energy Investment Allowance (EIA) to invest in a variety of energy-efficient and renewable energy technologies under favourable fiscal conditions, with available options listed in the annual Energy List of about 150 EIA-eligible investments. Companies can deduct 55.5 percent of the investment costs from the fiscal profits, on top of their usual depreciation. As a result, companies pay less income tax. On average, the EIA results in a 13.5 percent tax advantage. Alongside this tax advantage, the investments also ensure lower energy bills.[14]

Environmental Investment Rebate and Arbitrary Depreciation of Environmental Investments Scheme

The Environmental Investment Rebate (MIA), offering a tax refund on environmental investments, and the Arbitrary Depreciation of Environmental Investments (Vamil) scheme, offering voluntary depreciation on environmental investments, are two different schemes jointly managed by the Ministry of Infrastructure and the Environment and the Ministry of Finance. The aim of both schemes is to encourage Dutch entrepreneurs to invest in their business operations in an environmentally friendly way. The MIA scheme allows for the deduction of up to 36 percent of the cost of an environmentally friendly investment from fiscal profits while the Vamil scheme lets entrepreneurs decide when to write off 75 percent of their environmentally friendly investment costs and how fast or slow that will be. This, in turn, offers the entrepreneur an advantage in terms of liquidity and interest.

Investments that qualify for the MIA and Vamil schemes are found under the Environmental List, which contains 310 items of investment, or capital assets, that cause less environmental damage and are often in advance of the legal requirements. This means that if investments in these assets are made today the business will be ahead of their competitors when legislation and regulations are tightened. The Environmental List is updated annually with investments that fit less well in the MIA/Vamil schemes' objectives because of advancing technology removed and new innovative investments added, with the Ministry of Infrastructure and the Environment considering the latest policy topics such as biodiversity and climate change when making decisions on changes to the Environmental List. Businesses (suppliers and entrepreneurs) can propose a capital asset be included on the list. To be eligible, the environmental investment must at least

- Provide an obvious environmental benefit
- Be innovative or must still have a small market share in relation to the alternative
- Be more expensive than the environmentally unfriendly alternative[15]

Upstream Non-Fiscal Tools

The Netherlands has implemented a variety of upstream non-fiscal tools to develop the circular economy and encourage a life cycle perspective to be taken by economic actors in an attempt to decouple economic growth from resource use and associated environmental impacts.

Smart Regulation for Green Growth

The Smart Regulation for Green Growth programme is jointly managed by the Ministry of Economic Affairs and Ministry of Infrastructure and the Environment to find solutions to legislative and regulatory barriers encountered by businesses. Under the programme, if a business wishes to invest in a green innovation but finds itself restricted by legislation and regulations, then it can contact the programme's hotline and a multidisciplinary team will visit the business to discuss the issue and the underlying causes as well as consult with relevant stakeholders to find a solution. Each issue will be treated individually but the solutions will be shared. In some cases, it may be sufficient to review and explain the existing legislation to resolve the issue and facilitate the investment. In other cases, amendment of legislation is required, while at other times a technical amendment to regulations is sufficient. Overall, the aim of the programme is to create legal flexibility in favour of growth, investment, and innovation without losing sight of public values such as the environment. Information about resolved issues, descriptions of the relevant legislation, and amendments will be published on the programme's website.[16]

Voluntary Agreements

The Dutch consultative economy is based around people trying to further common interests through agreement, interchange, compromise, and cooperation. In the socioeconomic domain, this means that employers' and employees' organisations (the 'social partners') and government work together to shape national socioeconomic policy at the national level as well as at the sector and company level. The government plays an important role in the consultative economy as it represents the public interest and invites social partners to participate in developing socioeconomic policy. This participation may consist of making recommendations on critical issues or even result in agreements made at the national level that define the scope of those operating in sectors and within companies.[17]

The Energy Agreement for Sustainable Growth

The Energy Agreement for Sustainable Growth, concluded by the government with employers, trade unions, environmental organisations, and others, contains a series of provisions for achieving energy conservation,

Table 9.2 The Energy Agreement for Sustainable Growth provisions

Provision	Description
More Wind Turbines	1000 more wind turbines will be built. By 2020, 14 percent of all energy will be generated from renewable sources, rising to 16 percent in 2023
Resources for Home Insulation	The government is investing EUR 400 million in insulating rented homes. As well as reducing heating costs and CO^2 emissions, this measure will generate new jobs
More Jobs	Investments in energy from renewable sources and energy conservation will create 15,000 jobs
An Energy Label for Every Home	Since 2015, all privately owned and rented homes have been allocated a provisional energy label indicating the home's energy efficiency and raising awareness of energy consumption. This helps encourage people to invest in energy-saving measures
Tighter Agreements on Emissions Trading	The government wants the EU to improve the CO^2 emissions trading scheme and reduce emissions by at least 80 percent by 2050
National Energy Saving Fund	Homeowners can take out low-interest loans to fund energy-saving measures, financed by the National Energy Saving Fund, which has a budget of EUR 600 million
Tax Breaks for Local Clean Energy Initiatives	Local initiatives in which people collaborate to generate electricity from sustainable resources will be rewarded with lower energy tax rates for those involved

increasing energy from renewable sources, and job creation (summarised in Table 9.2). The governance and assurance of the Agreement is as follows:

- A permanent standing committee is created to operationalise general agreements, inspire and encourage parties to act, and monitor progress of actions and results.
- Each party is responsible for implementing its own actions.
- The parties are jointly responsible for the successful implementation of the Agreement.
- The committee monitors progress and will, when necessary, amend measures to achieve the targets.
- A dashboard will monitor progress of achieving the Agreement.
- An annual progress report along with a National Energy Report that assesses the policies will be published.[18]

Agreement on Sustainable Garment and Textile

In March 2016, a broad coalition of industry organisations, trade unions, civil society organisations, and the Dutch government created an agreement on international responsible business conduct in the garment and textile sector. The parties have agreed to work together when producing garments and textiles in countries such as Bangladesh, India, Pakistan, and Turkey to reduce the negative environmental impact of raw materials production, to use less water, energy, and chemicals, and to produce less chemical waste and wastewater (summarised in Table 9.3). This is in addition to preventing discrimination, child labour, and forced labour, promoting the right to collective bargaining by independent trade unions, ensuring living wages, providing healthy and safe working conditions, and preventing animal suffering.

The participating parties will work together on achieving sustainability targets that will be difficult or impossible to achieve alone. The businesses involved will draw up an annual action plan identifying specific goals, with the plan based on an investigation of problems and risks in their own production and supply chain. An independent secretariat, based at the Social and Economic Council of the Netherlands, will assess the quality and ambitions of these improvement plans. Meanwhile, unions and civil society organisations participating in the agreement will offer their knowledge and expertise and involve local partners in implementing the plans. Overall, Dutch consumers will gradually have access to fairer, more sustainable garments and textiles.[19]

The Nutrient Platform

The Nutrient Platform is a cross-sectoral network of Dutch organisations that believe in a pragmatic approach towards reducing nutrient scarcity. Thirty-six Dutch parties from the water sector, agriculture, waste sector, and chemical industry, including businesses, knowledge institutes, NGOs, and national ministries have joined forces to close nutrient cycles. The Nutrient Platform aims to accelerate the transition towards sustainable nutrient management by creating a market for recy-

Table 9.3 Agreement on Sustainable Garment and Textile resource use themes

Resource use	Aims	Measures to be taken by enterprises
Raw Material	To significantly reduce the environmental impact in the production or supply chain and create a circular economy in the garment and textile sector in the long-term	Make agreements in supply contracts to reduce environmental impact, combat wastage, and promote reuse Acquire knowledge of the environmental impact of different types of fibres and to carry out an analysis of the fibres used in their collections Set objectives or prepare an improvement plan to mitigate the impact of the total collection by using more environmentally friendly raw materials Establish criteria for sustainable use of raw materials and communicate them to suppliers and chain partners Join initiatives that make a direct contribution to reducing the environmental impact of raw materials
Water Pollution and Use of Chemicals, Water, and Energy	To significantly reduce the environmental impact caused by the use and discharge of water, energy, and chemicals in the production or supply chain	Make agreements in supply contracts on the prevention of pollution and inefficient use of raw materials in textile and garment production In countries where due diligence reveals there is a high risk of environmental pollution and inefficient use of raw materials, an additional emphasis is to be made on mapping the production or supply chain and building relationships with the stages (upstream in the chain) where textiles/garments undergo dyeing and finishing Actively request information from suppliers on the current situation and jointly identify options for improvement Acquire knowledge and implement trial projects to ascertain how a more aware design and product development process can minimise environmental and product safety risks Investigate whether existing certification schemes can help achieve targets Have a specific five-year strategy for achieving continuous improvement throughout the production or supply chain

SER. 2016. Agreement on Sustainable Garment and Textile. Available: http://www.ser.nl/en/publications/2016/agreement-sustainable-garment-textile.aspx

cled nutrients. Above all, the Nutrient Platform aims to turn the surplus of phosphorus in the Netherlands into an opportunity. This surplus, mainly the result of the Netherlands' intensive livestock system, causes environmental problems. By recovering phosphorus from 'waste' streams and turning it into valuable new products, not only is the environment improved but also the phosphorus cycle is closed. Overall, the Nutrient Platform will

- Develop innovative cross-sectoral business cases on nutrient recycling
- Improve Dutch and European legislation that helps turn 'waste' into resources
- Create awareness about sustainable nutrient management
- Link a knowledge agenda to business demand[20]

Downstream Fiscal Tools

The Netherlands has implemented a variety of downstream fiscal tools to develop the circular economy and encourage a life cycle perspective to be taken by economic actors in an attempt to decouple economic growth from resource use and associated environmental impacts.

Packaging Agreement for 2013–2022

In the Netherlands, producers and importers of packaged waste products are legally responsible for the prevention, collection, and recycling of packaging waste. Negotiations between Dutch industry, municipalities, and the national government on the responsibilities of producers and importers in the Netherlands resulted in the Packaging Agreement for 2013–2022. According to the Packaging Agreement, the following national recycling targets for packaging materials were to be reached by 2016:

- 46 percent of plastics
- 90 percent of glass

- 75 percent of paper
- 85 percent of metal
- 33 percent of wood

Each year after 2016, the targets for plastics and wood are raised by one percentage point and two percentage points respectively until 2022, when the targets will be fixed at 52 percent for plastics and 45 percent for wood.

Packaging Waste Fund

Producers and importers of more than 50,000 kg of packaging must register and pay a fee to the Packaging Waste Fund. The main aim of the fund is to cover the costs of implementing the Packaging Agreement. The fee for various types of waste is in Table 9.4 with the fee covering two components of the programme:

1. *Collection costs*: The costs associated with collecting and processing material and for acquiring a guarantee from recycling companies that they will use the collected materials

Table 9.4 Packaging Waste Fund fees

Material type	Rates 2016/2017 *EUR/kg*
Glass	0.0560
Paper Carton	0.0220
Plastic	0.6400
Bio-conductors	0.0200
Aluminium	0.0200
Other Metals	0.0200
Wood	0.0200
Other Formats	0.0200
Drinks Cartons	0.1800
Deposit Money Bottles	0.0200
Plastics, Non-Returnable	7.50

Afvalfonds. 2017b. *Rates* [Online]. Available: http://www.afvalfondsverpakkingen. nl/verpakkingen/alle-tarieven

2. *System costs*: The costs including the prevention of litter, monitoring, and organisational costs[21]

From Waste to Resource Programme

To government has developed the From Waste to Resource programme that connects knowledge and education to the circular economy. The main feature of the programme is that companies and research and educational institutions collaborate on circular designs, technologies, products, processes, and systems that enhance the circular economy. The programme also involves businesses and colleges cooperating on 110 projects related to the flows of materials including metal, rubber, plastic, and biotic waste streams.[22,23] A key programme is the Knowledge Innovation Mapping initiative.

Knowledge Innovation Mapping Funding

Under the Knowledge Innovation Mapping initiative, university researchers, in cooperation with companies, can apply for funding to conduct small, targeted circular economy projects that prevent waste and encourage high-quality reuse of materials. Research partners can apply for funding to conduct projects that focus on metals, plastics, rubber, nutrients, biotic materials, and waste streams. Overall, the programme promotes the knowledge of circular economy designs, technologies, processes, and system innovations with possible outcomes of projects including joint proposals, meetings, business models, demonstrator prototype(s), platforms, and tools. A maximum of EUR 20,000 is available for each 12-month-long project. In addition, participants must contribute to the project with EUR 5000 coming from companies involved and EUR 5000 from the university partner, resulting in total funding per project being EUR 30,000. All applications are assessed based on the following criteria:

- The contribution to closing the material chain
- Preventing waste and boosting resource savings
- Creating high-quality reuse materials[24]

Downstream Non-Fiscal Tools

The Netherlands has implemented a variety of downstream non-fiscal tools to develop the circular economy and encourage a life cycle perspective to be taken by economic actors in an attempt to decouple economic growth from resource use and associated environmental impacts.

Atlas of Natural Capital

The Atlas of Natural Capital provides information about natural capital and ecosystem services in the Netherlands. Users can see real-life examples of different ways of using the natural surroundings in a more sustainable and efficient way with the atlas making distinctions between different themes such as planning, water management, and nature management. Under each theme, there is a description of various subjects on how society can better use its natural surroundings. From there, users can click onwards to the geographical maps relevant to them. For individual users, they can find out how they can use natural capital more sustainably while companies can use this information to make their operations more sustainable and incorporate this information into their corporate social responsibility (CSR) reporting.[25]

Green Deal Circular Procurement

In 2013, the Green Deal Circular Procurement initiative, developed by CSR Netherlands, nevi, Rijkswaterstaat, and PIANOo, was launched to support entrepreneurs and the government in their ambitions of making circular purchases, where the purchaser ensures that the products or materials at the end of its life, service, or use can be used in a new cycle. Because this concept of circular purchasing is relatively new, participants share experiences and barriers, create momentum with new pilot projects, scale up ongoing circular initiatives, and integrate circular purchasing into participant organisations. The lessons learned are also disseminated to a broad audience so that other buyers do not need to 'reinvent the wheel'. Since 2013, all 45 members (public and private) have implemented at least two circular procurement processes. In 2017, members will come together in corporate meetings to develop ideas, solve problems, and achieve various innovations, with the following working groups established for the corporate meetings: construction, disposables, IT, infrastructure/civil engineering, interior, internal support, supply chain, and monitoring. In addition, throughout 2017, participants and their buyers will come together during four Community of Practice meetings to enhance knowledge of the circular economy and circular purchasing.[26]

Packaging Agreement 2013–2022: Plans for 2018 and 2022

The Packaging Agreement is an agreement in which industries should develop plans to ensure the sustainability of the packaging chain. The agreement is between the national government, municipalities, and producers and importers of packaged products. By 2013, all relevant industries had to develop a prevention plan for 2018 to obtain sustainability for the existing product/packaging combinations in the industry. In 2018 they have to develop a similar plan for 2022. The starting point for the plans are the highest obtainable objectives and goals which are determined by the Netherlands' Knowledge Institute Sustainable Packaging (KIDV). The plans will cover the following components:

- Packaging with improved recyclability
- Packaging with a higher ratio of recycled material
- Refill packaging
- Packaging that requires less raw materials
- Packaging of products which require less water or air use
- Packaging that extends the shelf life of a product
- Packaging that prevents product spoilage, damage, or spillage
- Packaging which causes less pollution by littering

The procedures for developing the plans are summarised as follows:

- The industries will develop sustainability plans and, based on data and insights of exemplary companies, determine the highest obtainable objectives for the industry, with the highest obtainable objectives compiled and determined by the following points:

  - *Reduce*: Preventing wastage of materials and more economical use of raw materials
  - *Reuse*: Increasing the percentage of reused packaging materials
  - *Recycle*: Recovery of packaging materials
  - *Renew*: Using new materials with reduced environmental pressure.

- Companies and industries can utilise the methodology developed by KIDV to determine the highest obtainable objectives. This is done by analysing the best-performing companies of each industry.
- In the approach used by KIDV, the highest obtainable objectives are logically based on the leading companies of the industry, resulting in a stimulus for the sustainability of the entire industry with frontrunners playing a prominent role.
- The KIDV reviews the highest obtainable objectives for each industry and determines these objectives.
- Upon request, the KIDV will support the leading companies on further steps such as research.[27]

CSR Netherlands

CSR Netherlands was founded in 2004 by the Ministry of Economic Affairs and has become a Centre of Excellence for Dutch companies, associations, NGOs, research institutions, healthcare providers, and public authorities to work together on translating CSR opportunities directly into market opportunities at three levels: the sector, the region, and the supply chain. Overall, CSR Netherlands shows the market opportunities that CSR offers, facilitates mutual collaboration, and provides useful

Table 9.5 CSR Netherlands' tools

Tool	Description
CSR Risk Check	This tool offers users insights into the issues that they may encounter in an organisation or company, starting from the earliest stages in the production chain. The risk check is easy to fill in and will instantly generate a list of possible issues for each country and/or product in the form of a PDF risk report
CSR MAPS	This online analysis program helps users determine the stage their company has reached in implementing CSR and what issues need to be prioritised. CSR MAPS will also show areas requiring improvement in the business operations and how the business can further develop their CSR policy
CSR STEPS	Many small and medium-sized enterprises (SMEs) need a clear roadmap to CSR. To help this group, CSR STEPS is an online tool that takes users through a step-by-step plan for establishing and implementing a CSR policy
CSR Guide International Organization for Standardization (ISO) 26000	This tool provides larger SMEs with an overview of how to integrate CSR into their company in accordance with ISO 26000 standards

information on applying CSR in practice. To support the business community with CSR, a variety of tools have been developed that are summarised in Table 9.5.[28,29]

Case Study Summary

The Netherlands' economy is the sixth-largest in Europe with its main economic sectors being wholesale and retail trade, transportation, food, and industry. Regarding agricultural production, the Netherlands is the second largest agricultural exporter in the world. While the economy is growing currently, it is projected to slow down in 2018. In addition, the Netherlands is experiencing a variety of challenges to its linear economy. Despite several air pollutants having decreased significantly in recent years, air pollution is still a major concern with many deaths linked to particulate matter. Climate change is likely to result in higher intensity extreme rainfall events during the winter months and droughts in the summer months, impacting water quantity and quality. In addition, infrastructure and people will be at risk of sea level rise. The Netherlands' power mix is dominated by natural gas, with the country being the second-largest producer and exporter of natural gas in Europe. Currently, the country is not on track to meet its renewable energy target in 2020. The country's population is growing rapidly due to an increase in migration and births, with most of the population growth concentrated in just four cities. Regarding waste, the Netherlands' municipal waste level is above the EU average. Finally, the main pressure on the country's water resources is diffuse pollution with nearly all water bodies affected.

Upstream

The Netherlands has implemented a variety of upstream fiscal and non-fiscal tools to develop the circular economy and encourage a life cycle perspective to be taken by economic actors in an attempt to decouple economic growth from resource use and associated environmental impacts. Some of the main tools are as follows:

- The Green Funds Scheme is a tax incentive that enables individual investors to invest in green projects. Green banking investments have a lower interest rate than the market rate and so this is compensated for by a tax incentive, with the size of the incentive dependent on the size of the investment made.
- The EIA encourages companies to invest in a range of energy-efficient technologies. Companies usually deduct a portion of their investment costs from the fiscal profits resulting in less income tax paid. In addition, companies often enjoy lower energy bills.
- The MIA offers a tax refund on environmental investments with companies able to deduct the cost of environmentally friendly investments from their fiscal profits while the Vamil scheme enables entrepreneurs to decide when to write off their environmentally friendly investment costs and how fast or slow that should be.

- The Smart Regulation for Green Growth programme provides support to businesses wishing to invest in circular economy innovations but find legislative and regulatory barriers. A multidisciplinary team will visit the business and consult relevant stakeholders to find a solution, which could include explaining existing legislation to resolve the issue or seek amendments to legislation or regulations.
- The Energy Agreement for Sustainable Growth, involving the government, employers, trade unions, and environmental organisations, seeks to achieve energy conservation, increase energy from renewable sources, and create jobs.
- The Agreement on Sustainable Garment and Textile, involving the government, industry organisations, trade unions, and civil society organisations, seeks to reduce the environmental impact of raw material use, encourage the reduction in water, energy, and chemicals, and reduce the amount of waste and wastewater in the garment and textile industry.
- The Nutrient Platform involves three-dozen Dutch parties from the water sector, agriculture, and waste sector along with government ministries joining forces to close nutrient cycles and create a market for recycled nutrients.

Downstream

The Netherlands has implemented a variety of downstream fiscal and non-fiscal tools to develop the circular economy and encourage a life cycle perspective to be taken by economic actors in an attempt to decouple economic growth from resource use and associated environmental impacts. Some of the main tools are as follows:

- Producers and importers of packaged waste products are responsible for the prevention, collection, and recycling of packaging waste. The Packaging Agreement contains a series of national recycling targets for packaging materials that were achieved by 2016. The Agreement also stipulates that relevant industries must develop packaging waste-prevention plans for 2018 and 2022.

- Knowledge Innovation Mapping funding is available for the conducting of small, targeted projects related to the circular economy with a specific focus on projects that prevent waste and encourage high-quality reuse of materials.
- The Packaging Waste Fund has been created to fund the costs of implementing the Packaging Agreement with all producers and importers creating a certain amount of packaging required to pay a fee per kilogram of material used.
- The Atlas of Natural Capital provides information about natural capital and ecosystem services in the Netherlands. Users can see real-life examples of how natural resources can be used in a more sustainable, efficient way. Companies can use the information to make their operations more sustainable and incorporate it in their CSR reporting.
- The Green Deal Circular Procurement initiative supports entrepreneurs and the government in making circular purchases where products or materials can be used in a new cycle. Throughout 2017, members of the initiative will come together in corporate meetings to develop new ideas, solve problems, and achieve various innovations with working groups established across a variety of sectors.
- CSR Netherlands, founded by a variety of partners from the government and civil society, aims to translate CSR opportunities directly into market opportunities at the sector, region, and supply chain levels. A variety of tools have been developed to provide insights into the types of CSR risks a business may face, to determine the stage they have reached in implementing CSR, and to help SMEs establish and implement a CSR policy.

Overall, the Netherlands has implemented a variety of upstream and downstream fiscal and non-fiscal tools to develop the circular economy and encourage a life cycle perspective to be taken by economic actors in an attempt to decouple economic growth from resource use and associated environmental impacts. These tools are summarised in Table 9.6.

Table 9.6 Netherlands case study summary

Tool	Tool type	Tool title	Description	Upstream/ Downstream
Fiscal	Environmental Taxes and Charges	Green Funds Scheme	A tax incentive that enables individual investors to invest in green projects	Upstream
		EIA	Encourages companies to invest in a range of energy-efficient technologies	Upstream
		MIA	Offers a tax refund on environmental investments with companies able to deduct the cost of environmentally friendly investments from their fiscal profits	Upstream
		Vamil	Enables entrepreneurs to decide when to write off their environmentally friendly investment costs	Upstream
		Packaging Waste Fund	All producers and importers creating a certain amount of packaging are required to pay a fee per kilogram of material used	Downstream
	Subsidies and Incentives	Knowledge Innovation Mapping Funding	Funding to conduct small, targeted circular economy projects that prevent waste and encourage reuse	Downstream

(continued)

Table 9.6 (continued)

Tool	Tool type	Tool title	Description	Upstream/ Downstream
Non-Fiscal	Regulations	Smart Regulation for Green Growth	Provides support to businesses wishing to invest in circular economy innovations but find legislative and regulatory barriers	Upstream
	Enhancing Business Competitiveness	From Waste to Resource Programme	Companies and research and educational institutions collaborate on circular economy designs, technologies, products, processes, and systems	Downstream
	Industry-Based Standards	The Nutrient Platform	Various stakeholders and the government have joined forces to close nutrient cycles and create a market for recycled nutrients	Upstream
		CSR Netherlands	Aims to translate CSR opportunities directly into market opportunities	Downstream
	Voluntary Agreements	Energy Agreement for Sustainable Growth	Seeks to achieve energy conservation, increase energy from renewable sources, and create jobs	Upstream
		Agreement on Sustainable Garment and Textile	Seeks to reduce the environmental impact of raw materials production and resource use in the international garment and textile industry	Upstream
		Packaging Agreement for 2013–2022	Contains a series of national recycling targets for packaging materials to be achieved and relevant industries must develop packaging waste-prevention plans	Downstream
	Greening the Supply Chain	Green Deal Circular Procurement	Supports entrepreneurs in making circular purchases where products or materials can be used in a new cycle	Downstream
	Information-Based Tools	Atlas of Natural Capital	Provides information about natural capital and ecosystem services in the Netherlands. Information can be inserted into CSR reports	Downstream

Notes

1. European Union. 2017. *Netherlands* [Online]. Available: https://europa.eu/european-union/about-eu/countries/member-countries/netherlands_en.
2. Holland Trade and Investment. 2017. *Export and import* [Online]. Available: https://www.hollandtradeandinvest.com/business-information/export--import.
3. European Commission. 2017a. *Economic forecast for the Netherlands* [Online]. Available: https://ec.europa.eu/info/business-economy-euro/economic-performance-and-forecasts/economic-performance-country/netherlands/economic-forecast-netherlands_en.
4. European Commission. 2017b. The EU environmental implementation review country report – The Netherlands. Available: http://ec.europa.eu/environment/eir/pdf/report_nl_en.pdf
5. Royal Netherlands Meteorological Institute. 2015. KNMI'14 climate scenarios. Climate scenarios for the Netherlands. Available: http://www.climatescenarios.nl/images/Brochure_KNMI14_EN_2015.pdf
6. ITA. 2016. *Netherlands – Energy* [Online]. Available: https://www.export.gov/article?id=Netherlands-Energy.
7. PBL Netherlands Environmental Assessment Agency. 2016. National energy outlook 2016. Available: http://www.pbl.nl/sites/default/files/cms/publicaties/pbl-2016-national-energy-outlook-2016.PDF.
8. CBS. 2017. *Relatively high population growth due to migration* [Online]. Available: https://www.cbs.nl/en-gb/news/2017/05/relatively-high-population-growth-due-to-migration.
9. NL Times. 2017. *Netherlands population crosses 17.1 million with immigration, birth increases* [Online]. Available: http://nltimes.nl/2017/01/03/netherlands-population-crosses-171-million-immigration-birth-increases.
10. European Commission. 2017b. The EU environmental implementation review country report – The Netherlands. Available: http://ec.europa.eu/environment/eir/pdf/report_nl_en.pdf.
11. Ibid.
12. Ministry of Housing, S. P. A. T. E. 2010. The green funds scheme. A success story in the making. Available: https://www.rvo.nl/sites/default/files/bijlagen/SEN040%20DOW%20A4%20Greenfunds_tcm24-119449.pdf.

13. Tax and Customs Administration. 2017. *Green investments* [Online]. Available: https://www.belastingdienst.nl/wps/wcm/connect/bldcontentnl/belastingdienst/prive/vermogen_en_aanmerkelijk_belang/vermogen/beleggen_met_belastingvoordeel/groene_beleggingen#.
14. Netherlands Enterprise Agency. 2017. *Energy investment allowance (EIA)* [Online]. Available: http://english.rvo.nl/subsidies-programmes/energy-investment-allowance-eia.
15. Netherlands Enterprise Agency. 2014. Tax relief schemes for environmentally friendly investment (Vamil and MIA). Available: http://english.rvo.nl/sites/default/files/2015/06/Introduction%20to%20MIA%20and%20Vamil%20-%20January%202014.pdf.
16. Ministry of Economic Affairs and Infrastructure and Environment. 2017. *Smart regulation* [Online]. Available: http://www.ruimteinregels.nl/english-smart-regulation/.
17. SER. 2015. The power of consultation. The Dutch consultative economy explained. Available: http://www.ser.nl/~/media/files/internet/talen/engels/brochure/informatiebrochure-power-consultation-en.ashx.
18. SER. 2014. Implementation of the Energy Agreement. Available: http://www.energieakkoordser.nl/~/media/files/energieakkoord/2014-implementation-energy-agreement.ashx.
19. SER. 2016. Agreement on Sustainable Garment and Textile. Available: http://www.ser.nl/en/publications/publications/2016/agreement-sustainable-garment-textile.aspx.
20. Nutrient Platform NL. 2017. *About nutrient platform* [Online]. Available: https://www.nutrientplatform.org/en/about-nutrient-platform/.
21. Afvalfonds. 2017a. *Legislative framework* [Online]. Available: http://www.afvalfondsverpakkingen.nl/en/legislative-framework.
22. Innovation Estafette. 2017. *Research and Training Program from Wastes to Raw Materials* [Online]. Available: https://www.innovatie-estafette.nl/onderwerpen/vang.
23. Regieorgaan-SIA. 2015. *Research program circular economics* [Online]. Available: http://www.regieorgaan-sia.nl/nieuws/pro-vang.
24. NWO. 2017. *Knowledge innovation mapping (KIEM) – From waste to raw material (VANG)* [Online]. Available: https://www.nwo.nl/financiering/onze-financieringsinstrumenten/sia/kiem-vang-kennisinnovatie-mapping-kiem---van-afval-naar-grondstof-vang/kiem-vang-kennisinnovatie-mapping-kiem---van-afval-naar-grondstof-vang.html.
25. Atlas Natural Capital. 2017. *About Atlas Natural Capital* [Online]. Available: http://www.atlasnatuurlijkkapitaal.nl/en/help.

26. MVO Nederland. 2017c. *Green deal* [Online]. Available: http://mvonederland.nl/green-deal-circulair-inkopen/green-deal.
27. Netherlands Institute for Sustainable Packaging. 2014. Establish highest obtainable objectives and sustainability plans for the industry. Available: https://www.kidv.nl/item/3821.
28. MVO Nederland. 2017a. *About us* [Online]. Available: http://mvonederland.nl/english/about-us/history.
29. MVO Nederland. 2017b. *CSR tools* [Online]. Available: http://mvonederland.nl/dossier/csr-tools.

References

Afvalfonds. 2017a. *Legislative framework* [Online]. Available: http://www.afvalfondsverpakkingen.nl/en/legislative-framework.
———. 2017b. *Rates* [Online]. Available: http://www.afvalfondsverpakkingen.nl/verpakkingen/alle-tarieven.
Atlas Natural Capital. 2017. *About atlas natural capital* [Online]. Available: http://www.atlasnatuurlijkkapitaal.nl/en/help.
CBS. 2017. *Relatively high population growth due to migration* [Online]. Available: https://www.cbs.nl/en-gb/news/2017/05/relatively-high-population-growth-due-to-migration.
European Commission. 2017a. *Economic forecast for the Netherlands* [Online]. Available: https://ec.europa.eu/info/business-economy-euro/economic-performance-and-forecasts/economic-performance-country/netherlands/economic-forecast-netherlands_en.
———. 2017b. The EU environmental implementation review country report – The Netherlands. Available: http://ec.europa.eu/environment/eir/pdf/report_nl_en.pdf.
European Union. 2017. *Netherlands* [Online]. Available: https://europa.eu/european-union/about-eu/countries/member-countries/netherlands_en.
Holland Trade and Investment. 2017. *Export and import* [Online]. Available: https://www.hollandtradeandinvest.com/business-information/export--import.
Innovation Estafette. 2017. *Research and training program from wastes to raw materials* [Online]. Available: https://www.innovatie-estafette.nl/onderwerpen/vang.
ITA. 2016. *Netherlands – Energy* [Online]. Available: https://www.export.gov/article?id=Netherlands-Energy.

Ministry of Economic Affairs and Infrastructure and Environment. 2017. *Smart regulation* [Online]. Available: http://www.ruimteinregels.nl/english-smart-regulation/

Ministry of Housing, S. P. A. T. E. 2010. The green funds scheme. A success story in the making. Available: https://www.rvo.nl/sites/default/files/bijla-gen/SEN040%20DOW%20A4%20Greenfunds_tcm24-119449.pdf.

MVO Nederland. 2017a. *About us* [Online]. Available: http://mvonederland.nl/english/about-us/history.

———. 2017b. CSR tools [Online]. Available: http://mvonederland.nl/dossier/csr-tools.

———. 2017c. *Green deal* [Online]. Available: http://mvonederland.nl/green-deal-circulair-inkopen/green-deal.

Netherlands Enterprise Agency. 2014. Tax relief schemes for environmentally friendly investment (Vamil and MIA). Available: http://english.rvo.nl/sites/default/files/2015/06/Introduction%20to%20MIA%20and%20Vamil%20-%20January%202014.pdf.

———. 2017. *Energy investment allowance* (*EIA*) [Online]. Available: http://english.rvo.nl/subsidies-programmes/energy-investment-allowance-eia.

Netherlands Institute for Sustainable Packaging. 2014. Establish highest obtainable objectives and sustainability plans for the industry. Available: https://www.kidv.nl/item/3821.

NL Times. 2017. *Netherlands population crosses 17.1 million with immigration, birth increases* [Online]. Available: http://nltimes.nl/2017/01/03/netherlands-population-crosses-171-million-immigration-birth-increases.

Nutrient Platform NL. 2017. *About nutrient platform* [Online]. Available: https://www.nutrientplatform.org/en/about-nutrient-platform/.

NWO. 2017. *Knowledge innovation mapping (KIEM) – From waste to raw material* (*VANG*) [Online]. Available: https://www.nwo.nl/financiering/onze-financieringsinstrumenten/sia/kiem-vang-kennisinnovatie-mapping-kiem---van-afval-naar-grondstof-vang/kiem-vang-kennisinnovatie-mapping-kiem---van-afval-naar-grondstof-vang.html.

PBL Netherlands Environmental Assessment Agency. 2016. National energy outlook 2016. Available: http://www.pbl.nl/sites/default/files/cms/publicaties/pbl-2016-national-energy-outlook-2016.PDF.

Regieorgaan-SIA. 2015. *Research program circular economics* [Online]. Available: http://www.regieorgaan-sia.nl/nieuws/pro-vang.

Royal Netherlands Meteorological Institute. 2015. KNMI'14 climate scenarios. Climate scenarios for the Netherlands. Available: http://www.climatescenarios.nl/images/Brochure_KNMI14_EN_2015.pdf.

SER. 2014. Implementation of the energy agreement. Available: http://www.energieakkoordser.nl/~/media/files/energieakkoord/2014-implementation-energy-agreement.ashx.

———. 2015. The power of consultation. The Dutch consultative economy explained. Available: http://www.ser.nl/~/media/files/internet/talen/engels/brochure/informatiebrochure-power-consultation-en.ashx.

———. 2016. Agreement on Sustainable Garment and Textile. Available: http://www.ser.nl/en/publications/publications/2016/agreement-sustainable-garment-textile.aspx.

Tax and Customs Administration. 2017. *Green investments* [Online]. Available: https://www.belastingdienst.nl/wps/wcm/connect/bldcontentnl/belastingdienst/prive/vermogen_en_aanmerkelijk_belang/vermogen/beleggen_met_belastingvoordeel/groene_beleggingen#.

10

Natural Resource Management and the Circular Economy in Scotland

Introduction

Scotland's GPB 117 billion gross value added economy is led by its top five export industries: food and drink, refined petroleum, business services, machinery and equipment, and electronic products. The economy only grew by 0.7 percent in the third quarter of 2016, compared to the UK's 2.2 percent. There are three reasons for this economic slowdown: first, the North Sea oil economy has declined due to a fall in oil prices in recent years, impacting the onshore economy; second, companies are shifting away from Scotland due a fear of political turmoil relating to independence and the European Union referendum resulting in businesses not being in the same regulatory zone as their customers; and third, a decline in public expenditure.[1]

Challenges to the Linear Economy

Scotland is experiencing a variety of challenges to its traditional linear economy as described below through an assortment of examples.

© The Author(s) 2018
R. C. Brears, *Natural Resource Management and the Circular Economy*,
Palgrave Studies in Natural Resource Management,
https://doi.org/10.1007/978-3-319-71888-0_10

Air Pollution

Air pollution in Scotland causes over 2500 premature deaths per annum. In urban centres, traffic is the dominant cause of air pollution, responsible for 80 percent of nitrogen dioxide pollution. In total, there are 39 declared pollution zones across the country. These zones are streets and areas where councils have declared that levels of air pollution are breaking standards set under the Scottish Air Quality Regulations, which are based on World Health Organization (WHO) guidelines. Even exposure to pollution levels that are lower than the regulatory limits is damaging to health if it persists over a long period. For instance, regular exposure to pollution can increase the risk of people with heart conditions having a heart attack or stroke, while new-born babies are likely to have lower birth rates.[2]

Climate Change

Climate change is likely to lead to regional summer temperatures increasing by 0.9–4.5 °C by the 2050s compared to the 1961–1990 baseline.[3] This will likely lead to a decrease in river flows and water availability during the summer, with some areas experiencing a decrease in runoff of up to 50 percent,[4] affecting water supplies for domestic users, industry, and business as well as the natural environment. Warmer, drier summers will change soil conditions, biodiversity, and landscape. Higher temperatures could lead to an increase in heat-related deaths in the summer: projections for the 2050s indicate around 100 additional premature deaths will occur, increasing to around 200 by the 2080s. Meanwhile, winter precipitation levels are projected to be up to 31 percent higher by the 2050s compared to the baseline, leading to more people and infrastructure being impacted from flooding events. It is projected that at least 40 percent more residential properties could be at a significant likelihood of flooding by the 2050s and at least 60 percent by the 2080s.[5,6,7]

Energy

Scotland is energy-rich, with oil and gas accounting for 87 percent of total primary energy in 2014. The result is that most of Scotland's heating and transport needs are supplied by fossil fuel. Nonetheless, the power sector has decarbonised with the closure of the last coal-fired power station in 2016. Already in 2015, renewables represented the largest source of electricity production (42 percent) followed by nuclear power (35 percent) and fossil fuel (22 percent).[8,9]

Population Growth

Over the 25-year period 2014–2039, Scotland's population is projected to rise by 7 percent from 5.35 million in 2014 to 5.70 million by 2039 and continue to rise into the future. The largest areas that will experience population increases will be the urban areas of Edinburgh (+21 percent) and Aberdeen (+17 percent).[10]

Waste

Scotland's quantity of Waste from All Sources (WFAS) was 11.63 million tonnes in 2015, an increase of 13.8 percent from 2014. This follows a 9.9 percent decrease in the previous year. The amount of waste generated varies between 10 and 20 percent year-on-year mainly due to the amount of construction and demolition (C&D) waste generated. This variability in C&D waste is sensitive to the number of large infrastructure projects in any given year.[11]

Water

The water supply-demand balance is highly uneven across Scotland with much of the demand for water in the drier east. Agricultural irrigation is increasing in many areas, which is linked to demand for higher-quality produce.[12]

Upstream Fiscal Tools

Scotland has implemented a variety of upstream fiscal tools to develop the circular economy and encourage a life cycle perspective to be taken by economic actors in an attempt to decouple economic growth from resource use and associated environmental impacts.

Resource Efficient Scotland Funding

Resource Efficient Scotland is a programme delivered by Zero Waste Scotland, a non-profit organisation funded by the Scottish Government, to help organisations save money by using resources more efficiently. Resource Efficient Scotland provides funding for a variety of circular economy initiatives.

Resource Efficient Scotland SME Loans

The Resource Efficient Scotland SME Loan aims to overcome financial barriers to implementing resource efficiency measures. The loan scheme supports small and medium-sized enterprises (SMEs), private-sector landlords, not-for-profit organisations, and charities in reducing costs through improved energy, material resource, and water efficiency by providing unsecured, interest-free loans between GBP 1000 and 100,000, repayable over a four-year period. The most common projects receiving loans include installation of new biomass boilers, upgrading to energy efficient lighting, upgrading to energy efficient heating, insulating premises, installing solar photovoltaic panels, upgrading of glazing, and optimising electricity voltages.[13]

Resource Efficiency Waste Prevention Implementation Fund

The Resource Efficiency Waste Prevention Implementation Fund is available to SMEs to support the physical implementation of waste prevention measures in the food and drink, construction, and other (commercial and industrial) sectors. The fund provides up to a maximum of GBP 100,000 per application for waste-prevention measures including

- Development of segregated recycling infrastructure areas
- Automated monitoring systems that detect and correct manufacturing faults that result in spoilt products
- Construction design to reduce end-of-life waste[14]

All applications will be evaluated against the following criteria and weightings before funding is granted:

- Return on investment (i.e. the value of support versus implemented savings) and economic impact on the business (35 percent)
- Environmental impact (35 percent)
- Other beneficial impacts (20 percent)
- Environmental commitment and overall commitment to health and safety (10 percent)[15]

Circular Economy Investment Fund

Zero Waste Scotland is investing GBP 18 million in the form of Circular Economy Investment Fund grants to help organisations create a more circular economy, with grant sizes ranging from a minimum of GBP 50,000 to a maximum of GBP 1 million. The fund invites applications from businesses and organisations working in all business and social economy sectors to

- Explore market feasibility for new circular economy products
- Develop and adopt innovative business models for new circular economy products and services
- Develop and uptake innovative technologies, products, and services to support a circular economy[16]

Resource Efficiency and Energy Efficiency Training Fund

Skills Development Scotland manages the Resource Efficiency and Energy Efficiency Training Fund that provides Scottish businesses with financial support for learning and training to improve resource and energy efficiency. The fund allows Scottish businesses with up to 250 employees to apply for a maximum of GBP 12,000 towards employee training, with the fund matching 50 percent of training costs up to a maximum of GBP 500 per session for a total of 25 training sessions. Training that is eligible for funding includes

- Renewable energy, low-carbon technologies, and microgeneration
- Energy efficiency, environmental, and clean technologies
- Waste management and reuse
- Reducing carbon in supply and energy management[17]

Low Energy Challenge Fund

The Low Energy Challenge Fund was launched in 2014 by the Scottish Government to support major demonstration projects that showcased transformative and innovative local energy solutions. Projects that received funding over the 2015/2016 period demonstrated partnership and collaborative learning, innovation, and added local value. The fund was a two-phase scheme with phase one consisting of a grant of up to GBP 30,000 to develop the project proposal, complete feasibility studies, and prepare for phase two, which was the delivery of the project over the 2015/2016 financial year. Successful phase two applicants received a maximum award of GBP 6 million per project. Due to the ambition for transformational projects, a minimum award threshold of GBP 1 million per project was set.[18]

Scottish Enterprise's Research and Development Grant

Scottish Enterprise's Research and Development (R&D) Grant supports businesses developing new products, processes, and services to improve company competitiveness and to benefit the Scottish economy. Projects typically last 6–36 months with grants covering up to 50 percent of the project costs. To be successful in receiving a grant, the project must represent a significant innovation for the company concerned and significant risks should be associated with the challenge of developing a new product, process, or service with project selection considering

- The nature of the R&D
- The creation or safeguarding of R&D jobs
- Linkages with other companies
- Global market opportunity
- Intellectual property
- The wider impact for society, including environmental impact, sustainability, and health and safety (grants will not be made to projects that have a known adverse effect on the environment and/or society)

The grant amount offered is at the discretion of Scottish Enterprise and ranges as follows:

- For all SMEs applying for a grant of excess of GBP 100,000, Scottish Enterprise will provide a grant of 35 percent up to 50 percent.
- For all large companies applying for a grant of more than GBP 100,000, Scottish Enterprise will provide a grant of 25 percent up to 40 percent.
- Grants above GBP 100,000 must demonstrate a positive impact on R&D jobs in Scotland.[19]

Upstream Non-Fiscal Tools

Scotland has implemented a variety of upstream non-fiscal tools to develop the circular economy and encourage a life cycle perspective to be taken by economic actors in an attempt to decouple economic growth from resource use and associated environmental impacts.

Green Champions Training

Resource Efficient Scotland has developed the Green Champions Training programme for Scottish SMEs to improve their resource efficiency and environmental performance. The programme is delivered in two parts: planning modules and practical modules. The planning modules introduce the threats and opportunities that are driving Scottish organisations to become more resource efficient before providing a quick step-by-step guide to making the participant's organisations more resource efficient. The practical modules provide training on a range of skills and techniques that will bring about long-term savings. Overall, the programme will provide participants with the skills to

- Understand the benefits of resource efficiency
- Identify opportunities to save money on energy, water, and materials
- Collect and analyse data
- Bring about change inside the organisation[20]

Circular Economy Business Support Service

Zero Waste Scotland's Circular Economy Business Support Service provides one-to-one support to small and medium-sized businesses to develop a more circular economy. The service is designed to help companies explore more circular ways of doing business which can result in resource efficiencies, improved profitability, higher-quality products, increased customer base, and alternative supply chains for businesses. The service is open to all eligible businesses to develop and/or implement new business models, technologies, practices, and products or services that can embed circular economy principles. Specific services include

- Innovative/opportunities identification
- Market assessment
- Stakeholder engagement
- Business plan support
- Life cycle analysis
- Business case development
- Commercial case development
- Support in identifying funding opportunities

- Communication and marketing support
- Implementation support[21]

One-to-One Support

Resource Efficient Scotland offers small and medium-sized businesses free on-site, one-to-one support to identify cost-saving opportunities. The support begins with an audit that lasts between two and three hours and allows an advisor to conduct an in-depth, structured review of the businesses' key operational areas and processes at their chosen site. After completion of the audit, the business will receive a report detailing the key actions that can be taken to unlock identified energy, water, and raw material cost savings. The report will also contain clear signposts on how to receive further advice, support, and funding to help implement the cost-saving actions effectively.[22]

Scottish Circular Economy Business Network

Zero Waste Scotland, in partnership with Scottish Enterprise, Highlands and Islands Enterprise, and the Scottish Environmental Protection Agency, has formed the Scottish Circular Economy Business Network (SCEBN). The SCEBN platform allows for engaged, innovative, and forward-thinking businesses to work collaboratively, share, and build on best practices to help Scotland transition towards a circular economy.[23]

Designing Out Construction Waste Guide

With the construction industry being Scotland's single largest waste producer, the Designing Out Construction Waste Guide highlights the opportunities to adopt a more circular approach to the construction process including waste efficient procurement, materials optimisation, off-site construction, reuse and recovery, deconstruction, and engaging with clients. Overall, the guide will help construction businesses reduce the environmental impact of construction at every stage of the build project; from design to procurement through to end-of-life deconstruction. It will also help make the best use of materials, water, and energy during construction and over the life cycle of built assets.[24]

Downstream Fiscal Tools

Scotland has implemented a variety of downstream fiscal tools to develop the circular economy and encourage a life cycle perspective to be taken by economic actors in an attempt to decouple economic growth from resource use and associated environmental impacts.

Carrier Bag Charge

The Carrier Bag Charge came into force in October 2014 and requires all retailers to charge a minimum of 5p for every single-use carrier bag. Zero Waste Scotland provides support and information to retailers to help them understand their responsibilities around the charge along with a dedicated website for retailers with a host of tools and materials including a staff training video.

Carrier Bag Commitment

Zero Waste Scotland operates Carrier Bag Commitment, which is a voluntary agreement that asks retailers to report carrier bag sales and commit to donating the money raised through the charge to good causes in Scotland, particularly ones that benefit the environment. By signing up to the commitment, retailers can access communication materials and contribute to national reporting on carrier bag use.[25,26]

Downstream Non-Fiscal Tools

Scotland has implemented a variety of downstream non-fiscal tools to develop the circular economy and encourage a life cycle perspective to be taken by economic actors in an attempt to decouple economic growth from resource use and associated environmental impacts.

Climate Change Assessment Tool

Resource Efficient Scotland has developed the Climate Change Assessment Tool (CCAT) to complement, inform, and support the public sector's existing climate change policies and practices. The CCAT is a self-assessment tool that enables public sector organisations to (1) assess their climate change credentials; (2) identify organisational strengths and weaknesses; (3) drive continuous improvement regarding carbon management; and (4) focus resources on key areas to improve sustainability, reduce emissions, and costs. The CCAT can be used up to five times enabling organisations to track their progress over time. There are five topics within CCAT with participants answering a total of 28 questions with results produced as percentage scores (higher percentage scores indicate the organisation is strong in that area). The five topics are summarised in Table 10.1.

Helping Businesses Measure and Monitor Resource Usage

Resource Efficient Scotland offers a wide array of tools to help businesses measure and monitor their resource usage. The tools are as follows.

Table 10.1 Climate Change Assessment Tool topics

Topic	Description
Governance	Governance refers to mechanisms, processes, and relations that direct and control organisations. Governance structures for climate change need to incorporate strategies and activities around mitigation and adaptation
Emissions	Emissions covers the carbon management aspect of climate change including having a strategy, plan, and targets to reduce emissions
Adaptation	Adaptation covers steps to manage climate risks
Behaviour	Behaviour refers to what staff understand about climate change and actions that can be taken to mitigate climate change
Procurement	Understanding how procurement of goods, services, and works can impact climate change

Resource Efficient Scotland. 2015. Climate change assessment tool. Available: http://www.resourceefficientscotland.com/sites/default/files/Guidance%20 Document%20for%20CCAT%20Tool.pdf

Measure and Monitor Water Use

The water usage tracking spreadsheet template helps businesses measure and monitor their water use. The template can be edited to suit the organisation and is accompanied by easy-to-follow instructions. In addition, businesses can watch a video on how to use the water usage tracking spreadsheet as well as organise a webinar on how to control water bills through effective measuring and monitoring.[27]

The Savings Finder

The Savings Finder provides businesses within minutes of using it online a free report highlighting the energy, water, and raw material cost savings that could be made. The report also includes an action plan that details step-by-step actions that can be taken to save money with signposts to the latest free tools, resources, and funding.[28]

Green Retail App

The Green Retail App provides a dedicated solution for all the resource efficiency needs of retailers, including videos, practical tools, and information to help the sector use energy, raw materials, water, and other resources as efficiently as possible. The smartphone app has been designed with off-the-shelf answers to help retailers save money by lowering operating costs and become more resource efficient. In addition, the app provides a link to free tailored advice and support on offer to all retailers in Scotland.[29]

Measure and Monitor Energy Use

Resource Efficient Scotland provides a template for businesses to measure and monitor the energy used in their organisation. This shows the true cost of waste and helps businesses focus their efforts on areas where the biggest cost savings can be made. The editable energy usage tracking spreadsheet helps businesses collate and analyse their measurement data. In addition, spreadsheet users can watch a video on how to use the measuring and monitoring spreadsheet as well as establish a webinar with Resource Efficient Scotland in which the organisation's energy efficiency experts will provide further tools and techniques to show businesses how they can measure and monitor energy use.[30]

Measure and Monitor Raw Materials and Waste

Businesses can use the raw materials and waste tracking spreadsheet to see the true cost of waste and help focus efforts on areas where the biggest cost savings can be made. The spreadsheet, which can be edited to suit the organisation, enables businesses to collate and analyse their measurement data. In addition, users can establish a webinar to see how a variety of tools and techniques can be used to measure and monitor raw materials and waste and reduce waste bills.[31]

Resource Efficiency Pledge

The Resource Efficiency Pledge is a nationwide scheme that helps Scottish businesses use energy, water, and raw materials more efficiently. By pledging to use resources more efficiently, businesses receive access to free, one-to-one advice and support tailored to the business. In addition, participating businesses can promote their commitment and differentiate themselves within the business community. The Pledge involves businesses signing up on a dedicated website and pledging to initiate between three and six actions or activities under the categories of Business Process, Supply Chain, Staff Engagement, and Custom over a 12-month period. Once actions have been achieved, the business is presented with a Bronze certificate. If the business decides to take further actions then they may be eligible for Silver, Gold, or even Platinum recognition, which is based on the amount of year-on-year sustainability commitments that they have made.[32]

Environmental Key Performance Indicators

Resource Efficient Scotland has developed a range of environmental key performance indicators (KPIs) to help businesses in a variety of sectors measure and monitor their resource use. The KPIs developed are summarised in Table 10.2.

Revolve Accreditation

Revolve Accreditation is Scotland's reuse quality standard for shops who sell second-hand goods with the quality standard being a sign of quality, reliability, and professionalism. To achieve certification, a shop must commit to continuously meeting high standards of customer care, shop layout, how they prepare goods for reuse, testing of goods, and health and safety. Specifically, the process involves a four-step process:

1. *Registration*: When a shop registers with the programme, they conduct a self-assessment based on the entry stage standards. Answering 'not met' to any of the statements does not preclude the shop from registering but gives an idea of the level of compliance required.
2. *Entry stage*: This involves checking if the organisation is legally compliant and has assessed all risks. This process consists of 43 standards covering trading standards, health and safety, product safety, governance, employment, and waste management. Passing this stage takes around two months. When an organisation starts this process, it will receive an entry stage assessment guide and a free health and safety audit from a specialist. Entry stage ends with an assessment visit from a case manager.
3. *Quality improvement stage*: The focus of this stage is on embedding a culture of professionalism and continuous improvement. This stage provides a broader assessment of the business and focuses on priority areas of improvement. This process takes around six months.
4. *Accreditation*: To achieve Revolve Accreditation the organisation must achieve both the following:

Table 10.2 Environmental key performance indicators

Type of business	Factors that drive resource use	KPI	KPI description
Food (cafes, bars, restaurants, canteens, and clubhouses)	Resource use will be driven by numbers of covers served. Number of hot drinks will also drive energy use	Energy	kWh (kilowatt hours) per cover served/hot drink sold
		Water	m³ per cover served
		Waste	Tonnes of recycling per cover served
			Tonnes of general/food waste per cover served
Hotels	Resource use will be driven by numbers of occupied rooms and number of guests	Energy	kWh per occupied room/guest
		Water	m³ per occupied room/guest
		Waste	Tonnes of recycling per occupied room/guest
			Tonnes of general waste per occupied room/guest
Manufacturers	Resource use will be driven by number of units produced and number of shifts	Energy	kWh per unit produced
			kWh per cycle/shift
		Water	m³ per unit produced
			m³ per cycle/shift
		Waste	Quantity of raw materials purchased per unit produced
			Tonnes of recycling per unit produced
			Tonnes of recycling per cycle/shift
			Tonnes of general waste per unit produced
			Tonnes of general waste per cycle/shift
Offices	Resource use will be driven by the number of staff working in the office	Energy	kWh per number of staff working in the office
		Water	m³ per number of staff working in the office
		Waste	Tonnes of recycling per number of staff working in the office
			Tonnes of general waste per number of staff working in the office

(continued)

Table 10.2 (continued)

Type of business	Factors that drive resource use	KPI	KPI description
Retailers	Energy use will be driven by the number of hours open. Water use will also be driven by the number of hours open and the number of staff working	Energy	kWh per number of hours open
		Water	m³ per number of hours open/staff working
		Waste	Tonnes of recycling per volume of deliveries
			Tonnes of general waste per volume of deliveries
Entertainment Venues	Resource use will be driven by footfall (attendee numbers) and number of showings/events	Energy	kWh per footfall/number of attendees
			kWh per number of showings/events
		Water	m³ per footfall/number of attendees
			m³ per number of showings/events
		Waste	Tonnes of recycling per footfall/number of attendees
			Tonnes of recycling per number of showings/events
			Tonnes of general waste per footfall/number of attendees
			Tonnes of general waste per number of showings/events

Resource Efficient Scotland. 2016b. Try these environmental KPIs. Available: http://www.resourceefficientscotland.com/ sites/default/files/Try%20these%20environmental%20KPIs.pdf

- The European Foundation for Quality Management 'Committed to Excellence' award, which is validated by Quality Scotland. This award helps the organisation develop and integrate a culture of continuous improvement. It focuses on priority areas of performance that are relevant for reuse businesses.
- Participation in the Revolve retail programme and pass the Revolve retail audit with the retail programme consisting of training and professional support to help maximise business potential.[33]

Measuring to Manage Your Resources Implementation Guide

Resource Efficient Scotland has developed the Measuring to Manage Your Resources Implementation Guide that takes businesses through a stage-by-stage process of using resources more efficiently. The guide explains techniques and provides links to useful tools and further support along the way. The guide is broken down into five stages summarised in Table 10.3.

Mandatory Commercial Recycling

Since 1 January 2014, it has been mandatory for all businesses and organisations operating in Scotland, regardless of size, to recycle plastics, metals, paper, card, and glass. Food businesses (producers, preparers, or sellers of food) which produce over 50 kg of food waste per week must present it for separate collection unless they are in a rural area. To help businesses prepare for the new regulations Zero Waste Scotland, via the Resource Efficient Scotland website, offered a variety of user-friendly, informative resources to help make the transition to the new regulations as smooth as possible. The resources included:

- Video case studies which demonstrated how organisations of different sizes in a variety of sectors have put systems in place and trained staff in preparation for the new rules

Table 10.3 Measuring to Manage Your Resources Implementation Guide stages

Stage	Action	Description
1	Understanding what is driving resource use	Readers identify the factors that are driving how much waste the business is producing and how much energy and water is used. From this, KPIs can be selected that will help businesses understand the performance of their organisation and identify opportunities for improvement
2	Identifying data requirements	Once KPIs have been identified and benchmarks established for comparison, readers then understand what types of data need collecting
3	Collecting data	Readers understand the processes of measuring and monitoring the data. This will enable businesses to not only develop a historical picture of past resource consumption but also enable them to start identifying performance issues and immediate savings opportunities
4	Analysing and evaluating the data	The next stage is to analyse and evaluate the data to identify any problems and identify areas for possible improvements and cost savings. This can be done at the company, site, department, or process level depending on the data collected
5	Identifying improvements and acting	Businesses can prepare a list of actions to address opportunities that have been identified. To help do this, Resource Efficient Scotland has developed a range of free guides that provide advice on how to save money and energy from space heating and lighting, as well as save money on waste and water usage

Resource Efficient Scotland. 2016a. Measuring to manage your resources. Available: http://www.resourceefficientscotland.com/sites/default/files/Measuring%20to%20manage%20your%20resources.pdf

- A frequently asked questions database that provided instant answers to common questions on waste management
- A rural postcode finder to help businesses determine whether they are required to separate food waste under the regulations

- A poster creator tool that enabled businesses to develop clear signage for recycling bins to ensure employees understood what and where to recycle
- A business reuse and recycling directory that helped businesses easily source waste management services including recycling and reuse organisations operating in their postal code[34]

Supply Chain Management and Sustainable Procurement: A Guide for Scottish SMEs

The Supply Chain Management and Sustainable Procurement: A Guide for Scottish SMEs provides practical advice and guidance about supply chain management and the role SMEs have in influencing their customers in a positive way on the goods and services they buy and use. The guide explains a variety of techniques for assessing the environmental impact of the goods and services supplied and purchased. Practical steps are offered that can improve the sustainability of the supply chain. The guide also recommends a partnership approach enabling businesses and customers to work together in reducing costs, increasing efficiency, and reducing joint environmental impacts.[35]

Site Waste Management Plan

Resource Efficient Scotland has created a free online tool to manage waste on construction sites by tracking materials and waste. The Site Waste Management Plan tool provides a template to help companies:

- Evaluate their total waste by category, for example, glass, concrete, floor tiles, and carpets, and break down how much is reused or recycled, both on-site and off, and how much is landfilled.
- Identify opportunities to reduce waste and evaluate the merits of retrofits, refurbishments, and fitting and stripping-out projects.
- Create an audit trail to prove compliance with waste regulations.[36]

Case Study Summary

Scotland's economy is export-driven with the largest sectors being food and drink, petroleum, services, industry, and electronic products. Economic growth is significantly below the UK's level due to weakening oil prices, political instability, as well as a decline in public expenditure. Scotland is experiencing a variety of challenges to its linear economy. There are many regions across Scotland that have been declared Pollution Zones, which are areas that break air quality standards based on WHO guidelines. With climate change, rising temperatures will see areas of Scotland experiencing a decrease in water availability that will impact all water users, human and natural. Meanwhile, there will be higher levels of precipitation during the winter months, exposing people and infrastructure to increased flooding risks. As Scotland is energy-rich, oil and gas are a key source of energy, with most of Scotland's heating and transport needs supplied by fossil fuel. Over the next several decades, Scotland's population is projected to rise with the largest increases expected to be in Edinburgh and Aberdeen. Scotland's quantity of waste from all sources has increased in recent years, following a decline in the preceding year. This is due to variability in C&D waste generated each year. Finally, the water supply-demand balance is highly uneven across Scotland with high demand in the drier east.

Upstream

Scotland has implemented a variety of upstream fiscal and non-fiscal tools to develop the circular economy and encourage a life cycle perspective to be taken by economic actors in an attempt to decouple economic growth from resource use and associated environmental impacts. Some of the main tools are as follows:

- Resource Efficient Scotland is offering SME loans to organisations to overcome financial barriers to implementing resource efficiency measures. The loans enable improved energy, material resource, and water efficiency.

- The Resource Efficiency Waste Prevention Implementation Fund is available to SMEs to support the physical implementation of waste prevention measures.
- The Circular Economy Investment Fund is providing grants to small and medium-sized businesses that are helping create a more circular economy. Grant applications are being sought from businesses and organisations developing new circular economy products, services, and business models.
- Skills Development Scotland is providing financial support for learning and training initiatives that improve resource and energy efficiency.
- The Low Energy Challenge Fund was launched to support major demonstration projects that provided transformative and innovative local energy solutions. Projects that received funding should have demonstrated partnership and collaborative learning as well as be innovative and add local value.
- Scottish Enterprise's R&D Grant supports businesses developing new products, processes, and services that improve company competitiveness and benefit the Scottish economy. Grants will be awarded to projects that consider the wider impact for society including environmental impact.
- The Green Champions Training programme for Scottish SMEs helps improve resource efficiency and environmental performance.
- The Circular Economy Business Support Service provides one-to-one support to small and medium-sized businesses to develop a more circular economy.
- Resource Efficient Scotland offers one-to-one support to small and medium-sized businesses to help them identify resource efficiency-related cost-saving opportunities.
- The Circular Economy Business Network has been formed to encourage innovative, forward-thinking businesses to work collaboratively, share, and build best practices to help Scotland transition to the circular economy.
- The Designing Out Construction Waste Guide highlights the opportunities of designing out waste across the entire construction process.

Downstream

Scotland has implemented a variety of downstream fiscal and non-fiscal tools to develop the circular economy and encourage a life cycle perspective to be taken by economic actors in an attempt to decouple economic growth from resource use and associated environmental impacts. Some of the main tools are as follows:

- The Carrier Bag Charge requires all retailers charge for single-use carrier bags. As part of the initiative, the Carrier Bag Commitment voluntary agreement asks retailers to report carrier bag sales and commit to donating the money raised to good causes in Scotland.
- The CCAT is designed to complement, inform, and support existing climate change policies and practices of the public sector including in procurement.
- Resource Efficient Scotland provides a water usage tracking spreadsheet template to help businesses analyse their water usage. The template comes with easy-to-follow instructions, a video on how to use the spreadsheet, and the opportunity to watch a webinar on how to control water bills.
- The Savings Finder is a free online tool that provides businesses with a report on where to find energy, water, and raw material cost savings. The report includes step-by-step actions that can be taken to save money.
- The Green Retail App provides retailers with tools and information to use energy, raw materials, water, and other resources as efficiently as possible. The app also provides a link to free tailored advice and support on offer to all retailers in Scotland.
- Resource Efficient Scotland provides businesses with a template to measure and monitor their energy usage. This helps businesses focus their efforts on where the biggest cost savings can be made. A webinar also shows how businesses can measure and monitor energy use.
- Businesses can use the raw materials and waste tracking spreadsheet to show the true cost of waste and help focus efforts on areas where the

biggest cost savings can be made. In addition, users can watch a webinar to control waste bills.

- The Resource Efficiency Pledge helps Scottish businesses use resources more efficiently. By making a pledge, businesses can access free, one-to-one advice and support tailored to their business. Participating businesses also receive recognition awards.
- Resource Efficient Scotland has developed a range of environmental KPIs to help businesses in a variety of sectors measure and monitor their resource usage.
- Revolve Accreditation is Scotland's reuse quality standard for shops that sell second-hand goods with the quality standard being a sign of quality, reliability, and professionalism. To receive certification, a shop must commit to meeting a variety of high standards.
- Resource Efficient Scotland has developed the Measuring to Manage Your Resources Implementation Guide for businesses to help them measure and manage their resource consumption.
- Scotland has mandated that all businesses and organisations must recycle a range of materials including plastics, paper, and glass. Meanwhile, food businesses which produce a certain amount of waste or more per week must present it for separate collection.
- The Supply Chain Management and Sustainable Procurement: A Guide for Scottish SMEs provides practical advice and guidance about supply chain management and the role SMEs have in influencing their customers in a positive way about the goods and services they buy and use.
- The Site Waste Management Plan tool helps companies manage waste on construction sites by tracking materials and waste. The tool evaluates total waste by category and how much is reused or recycled as well as identifies opportunities to reduce waste.

Overall, Scotland has implemented a variety of upstream and downstream fiscal and non-fiscal tools to develop the circular economy and encourage a life cycle perspective to be taken by economic actors in an attempt to decouple economic growth from resource use and associated environmental impacts. These tools are summarised in Table 10.4.

Table 10.4 Scotland case study summary

Tool	Tool type	Tool title	Description	Upstream/Downstream
Fiscal	Environmental Taxes and Charges	Carrier Bag Charge	Requires all retailers to charge for single-use carrier bags	Downstream
	Subsidies and Incentives	Resource Efficient Scotland SME Loan	Loans to overcome financial barriers to implementing resource efficiency measures	Upstream
		Resource Efficiency Waste Prevention Implementation Fund	Available to SMEs to support the physical implementation of waste-prevention measure	Upstream
		Circular Economy Investment Fund	Grants to small and medium-sized businesses who are helping create a more circular economy	Upstream
		Resource Efficiency and Energy Efficiency Training Fund	Financial support for learning and training initiatives that improve resource and energy efficiency	Upstream
		Low Energy Challenge Fund	Supported major demonstration projects that provided transformative and innovative local energy solutions	Upstream
		Scottish Enterprise's R&D Grant	Supports businesses developing new products, processes, and services that improve competitiveness	Upstream
Non-Fiscal	Regulations	Mandatory Commercial Recycling	All businesses and organisations operating in Scotland must recycle a range of materials. Some food businesses must separate their food waste	Downstream
	Green Public Procurement	CCAT	Designed to complement, inform, and support existing climate change policies and practices of the public sector	Downstream

(continued)

Table 10.4 (continued)

Tool	Tool type	Tool title	Description	Upstream/ Downstream
	Enhancing Business Competitiveness	KPIs	Helps businesses in a variety of sectors measure and monitor their resource usage	Downstream
	Education and Training	Green Champions Training	Training programme for Scottish SMEs to help improve resource efficiency and environmental performance	Upstream
	Raising Industry Awareness and Capacity	Circular Economy Business Support Service	Provides one-to-one support to small and medium-sized businesses to develop a more circular economy	Upstream
		One-to-One Support	Offers small and medium-sized businesses free on-site, one-to-one support to identify cost-saving opportunities	Upstream
		Designing Out Construction Waste Guide	Highlights the opportunities of designing out waste across the entire construction process	Upstream
		Measuring to Manage Your Resources Implementation Guide	Helps businesses measure and manage their resource consumption	Downstream
	Eco-Labels and Certification	Revolve Accreditation	A quality standard for shops that sell second-hand goods	Downstream
	Greening the Supply Chain	Supply Chain Management and Sustainable Procurement: A Guide for Scottish SMEs	Provides practical advice and guidance to SMEs about supply chain management and the role they have in positively influencing their customers	Downstream
	Environmental Recognition Awards	Resource Efficiency Pledge	Pledging businesses can access free, one-to-one advice and tailored support as well as receive recognition awards for actions taken	Downstream

(continued)

Table 10.4 (continued)

Tool	Tool type	Tool title	Description	Upstream/Downstream
	Knowledge Transfer Networks	Circular Economy Business Network	Formed to encourage innovative businesses to work collaboratively, share, and build a circular economy	Upstream
	Information-Based Tools	Water Usage Tracking	The spreadsheet template helps businesses analyse their water usage	Downstream
		Savings Finder	A free, online tool that provides businesses with a report on where to find resources and raw material cost savings	Downstream
		Green Retail App	Provides retailers with tools and information to use resources as efficiently as possible	Downstream
		Measure and Monitor Energy Usage	Businesses are provided with a template and webinar to measure and monitor their energy usage	Downstream
		Measure and Monitor Raw Materials and Waste	Businesses can use the waste tracking spreadsheet and watch a webinar to focus efforts on areas where the biggest cost savings can be made	Downstream
		Site Waste Management Plan	Helps companies manage waste on construction sites	Downstream

Notes

1. CEBR. 2017. *Scotland has a growth problem and a deficit problem – But over time they can be fixed* [Online]. Available: https://cebr.com/reports/scotland-has-a-growth-problem-and-a-deficit-problem-but-over-time-they-can-be-fixed/.
2. Friends of the Earth Scotland. 2017. *Air pollution* [Online]. Available: http://www.foe-scotland.org.uk/air-pollution.
3. Committee on Climate Change. 2017. UK climate change risk assessment 2017. Available: http://www.adaptationscotland.org.uk/application/files/4314/7792/6358/UK-CCRA-2017-Scotland-National-Summary.pdf.
4. CREW. 2012. The water supply-demand balance and climate change. Available: http://www.crew.ac.uk/sites/default/files/sites/default/files/publication/Climate_change_and_water_demand-supply_summary.pdf.
5. Scottish Government. 2017b. *Preparing for a new climate* [Online]. Available: http://www.gov.scot/Topics/Environment/climatechange/scotlands-action/adaptation.
6. SEPA. 2017. *Water scarcity* [Online]. Available: https://www.sepa.org.uk/environment/water/water-scarcity/
7. Committee on Climate Change. 2017. UK climate change risk assessment 2017. Available: http://www.adaptationscotland.org.uk/application/files/4314/7792/6358/UK-CCRA-2017-Scotland-National-Summary.pdf.
8. Scottish Government. 2017a. Draft Scottish energy strategy: The future of energy in Scotland. Available: http://www.gov.scot/Publications/2017/01/3414.
9. Scottish Government. 2016a. *High level summary of statistics trend last update: Thursday, December 22, 2016. Electricity generation* [Online]. Available: http://www.gov.scot/Topics/Statistics/Browse/Business/TrendElectricity.
10. Scottish Government. 2016b. *Population projections for Scottish areas* [Online]. Available: https://news.gov.scot/news/population-projections-for-scottish-areas-2.
11. SEPA. 2015. Waste from all sources – Summary data for 2015. Available: https://www.sepa.org.uk/media/287063/waste-from-all-sources-summary-data-2015.pdf

12. CREW. 2012. The water supply-demand balance and climate change. Available: http://www.crew.ac.uk/sites/default/files/sites/default/files/publication/Climate_change_and_water_demand-supply_summary.pdf.

13. Government of Scotland. 2017. Resource Efficient Scotland – SME loans. Available: https://www.mygov.scot/resource-efficient-scotland-sme-loans/?via=http://business.scotland.gov.uk/view/funding/resource-efficient-scotland-sme-loans.

14. Resource Efficient Scotland. 2017j. *Waste prevention implementation fund* [Online]. Available: http://www.resourceefficientscotland.com/resource/resource-efficiency-waste-prevention-implementation-fund.

15. Zero Waste Scotland. 2015. Guidance document. Available: http://www.zerowastescotland.org.uk/sites/default/files/4RE003-000%20Guidance%20document%20FINAL_2.pdf.

16. Zero Waste Scotland. 2017c. *Circular economy investment fund* [Online]. Available: http://www.zerowastescotland.org.uk/circular-economy/investment-fund.

17. Government of Scotland. 2014. Agri-renewables strategy for Scotland. Available: http://www.gov.scot/resource/0044/00443422.pdf.

18. Local Energy Scotland. 2017. *The local energy challenge fund* [Online]. Available: http://www.localenergyscotland.org/funding-resources/funding/local-energy-challenge-fund/.

19. Scottish Enterprise. 2017. *Finance your research* [Online]. Available: https://www.scottish-enterprise.com/services/develop-new-products-and-services/rd-grant/overview.

20. Resource Efficient Scotland. 2017c. *Green champions training* [Online]. Available: http://www.resourceefficientscotland.com/resource/green-champions-training.

21. Zero Waste Scotland. 2017b. *Circular economy business support service* [Online]. Available: http://www.zerowastescotland.org.uk/circular-economy/business-support-service.

22. Resource Efficient Scotland. 2017g. *One-to-one support* [Online]. Available: http://www.resourceefficientscotland.com/content/one-one-support.

23. Zero Waste Scotland. 2017f. *Scottish circular economy business network* [Online]. Available: http://www.zerowastescotland.org.uk/circular-economy/scottish-network.

24. Resource Efficient Scotland. 2017b. *Design out waste in construction* [Online]. Available: http://www.resourceefficientscotland.com/resource/designing-out-construction-waste-guide.

25. Zero Waste Scotland. 2017a. *Carrier bag charge* [Online]. Available: http://www.zerowastescotland.org.uk/litter-flytipping/carrier-bag-charge.

26. Zero Waste Scotland. 2017d. *Meeting regulations and legislation* [Online]. Available: http://www.zerowastescotland.org.uk/content/meeting-regulations-legislation.

27. Resource Efficient Scotland. 2017f. *Measure and monitor your water use* [Online]. Available: http://www.resourceefficientscotland.com/content/key-task/measure-monitor-waste-water-use.

28. Resource Efficient Scotland. 2017i. *The savings finder* [Online]. Available: http://www.resourceefficientscotland.com/resource/savings-finder.

29. Zero Waste Scotland. 2017e. *New app makes going green easy for retailers* [Online]. Available: http://www.zerowastescotland.org.uk/content/new-app-makes-going-green-easy-retailers.

30. Resource Efficient Scotland. 2017d. *Measure and monitor energy use* [Online]. Available: http://www.resourceefficientscotland.com/resource/measure-and-monitor-energy-use.

31. Resource Efficient Scotland. 2017e. *Measure and monitor your business waste* [Online]. Available: http://www.resourceefficientscotland.com/resource/measure-and-monitor-raw-materials-and-waste.

32. Resource Efficient Scotland. 2017h. *Resource efficient pledge* [Online]. Available: http://www.resourceefficientscotland.com/resource/resource-efficiency-pledge.

33. Revolve. 2017. *Revolve re-use quality standard* [Online]. Available: http://www.revolvereuse.com/.

34. Recycle for Scotland. 2017. *Recycle at work* [Online]. Available: http://www.recycleforscotland.com/recycle/recycle-work.

35. Zero Waste Scotland. 2011. Supply chain management and sustainable procurement. A guide for Scottish SMEs. Available: http://www.resourceefficientscotland.com/sites/default/files/Supply%20Chain%20Management%20and%20Sustainable%20Procurement%20-%20A%20guide%20for%20Scottish%20SMEs.pdf.

36. Resource Efficient Scotland. 2017a. *Create a site waste management plan with our help* [Online]. Available: http://www.resourceefficientscotland.com/Construction/waste-management-plan.

References

CEBR. 2017. *Scotland has a growth problem and a deficit problem – But over time they can be fixed* [Online]. Available: https://cebr.com/reports/scotland-has-a-growth-problem-and-a-deficit-problem-but-over-time-they-can-be-fixed/.

Committee on Climate Change. 2017. UK climate change risk assessment 2017. Available: http://www.adaptationscotland.org.uk/application/files/4314/7792/6358/UK-CCRA-2017-Scotland-National-Summary.pdf.

CREW. 2012. The water supply-demand balance and climate change. Available: http://www.crew.ac.uk/sites/default/files/sites/default/files/publication/Climate_change_and_water_demand-supply_summary.pdf.

Friends of the Earth Scotland. 2017. *Air pollution* [Online]. Available: http://www.foe-scotland.org.uk/air-pollution.

Government of Scotland. 2014. Agri-renewables strategy for Scotland. Available: http://www.gov.scot/resource/0044/00443422.pdf.

———. 2017. Resource efficient Scotland – SME loans. Available: https://www.mygov.scot/resource-efficient-scotland-sme-loans/?via=http://business.scotland.gov.uk/view/funding/resource-efficient-scotland-sme-loans.

Local Energy Scotland. 2017. *The local energy challenge fund* [Online]. Available: http://www.localenergyscotland.org/funding-resources/funding/local-energy-challenge-fund/.

Recycle for Scotland. 2017. *Recycle at work* [Online]. Available: http://www.recycleforscotland.com/recycle/recycle-work.

Resource Efficient Scotland. 2015. Climate change assessment tool. Available: http://www.resourceefficientscotland.com/sites/default/files/Guidance%20Document%20for%20CCAT%20Tool.pdf.

———. 2016a. Measuring to manage your resources. Available: http://www.resourceefficientscotland.com/sites/default/files/Measuring%20to%20manage%20your%20resources.pdf.

———. 2016b. Try these environmental KPIs. Available: http://www.resourceefficientscotland.com/sites/default/files/Try%20these%20environmental%20KPIs.pdf.

———. 2017a. *Create a site waste management plan with our help* [Online]. Available: http://www.resourceefficientscotland.com/Construction/waste-management-plan.

———. 2017b. *Design out waste in construction* [Online]. Available: http://www.resourceefficientscotland.com/resource/designing-out-construction-waste-guide.

————. 2017c. *Green champions training* [Online]. Available: http://www.resourceefficientscotland.com/resource/green-champions-training.

————. 2017d. *Measure and monitor energy use* [Online]. Available: http://www.resourceefficientscotland.com/resource/measure-and-monitor-energy-use.

————. 2017e. *Measure and monitor your business waste* [Online]. Available: http://www.resourceefficientscotland.com/resource/measure-and-monitor-raw-materials-and-waste.

————. 2017f. *Measure and monitor your water use* [Online]. Available: http://www.resourceefficientscotland.com/content/key-task/measure-monitor-waste-water-use.

————. 2017g. *One-to-one support* [Online]. Available: http://www.resourceefficientscotland.com/content/one-one-support.

————. 2017h. *Resource efficient pledge* [Online]. Available: http://www.resourceefficientscotland.com/resource/resource-efficiency-pledge.

————. 2017i. *The savings finder* [Online]. Available: http://www.resourceefficientscotland.com/resource/savings-finder.

————. 2017j. *Waste prevention implementation fund* [Online]. Available: http://www.resourceefficientscotland.com/resource/resource-efficiency-waste-prevention-implementation-fund.

Revolve. 2017. *Revolve re-use quality standard* [Online]. Available: http://www.revolvereuse.com/.

Scottish Enterprise. 2017. *Finance your research* [Online]. Available: https://www.scottish-enterprise.com/services/develop-new-products-and-services/rd-grant/overview.

Scottish Government. 2016a. *High level summary of statistics trend last update: Thursday, December 22, 2016. Electricity generation* [Online]. Available: http://www.gov.scot/Topics/Statistics/Browse/Business/TrendElectricity.

————. 2016b. *Population projections for Scottish areas* [Online]. Available: https://news.gov.scot/news/population-projections-for-scottish-areas-2.

————. 2017a. Draft Scottish energy strategy: The future of energy in Scotland. Available: http://www.gov.scot/Publications/2017/01/3414.

————. 2017b. *Preparing for a new climate* [Online]. Available: http://www.gov.scot/Topics/Environment/climatechange/scotlands-action/adaptation.

SEPA. 2015. Waste from all sources – Summary data for 2015. Available: https://www.sepa.org.uk/media/287063/waste-from-all-sources-summary-data-2015.pdf.

———. 2017. *Water scarcity* [Online]. Available: https://www.sepa.org.uk/environment/water/water-scarcity/.

Zero Waste Scotland. 2011. Supply chain management and sustainable procurement. A guide for Scottish SMEs. Available: http://www.resourceefficientscotland.com/sites/default/files/Supply%20Chain%20Management%20and%20Sustainable%20Procurement%20-%20A%20guide%20for%20Scottish%20SMEs.pdf.

———. 2015. Guidance document. Available: http://www.zerowastescotland.org.uk/sites/default/files/4RE003-000%20Guidance%20document%20FINAL_2.pdf.

———. 2017a. *Carrier bag charge* [Online]. Available: http://www.zerowastescotland.org.uk/litter-flytipping/carrier-bag-charge.

———. 2017b. *Circular economy business support service* [Online]. Available: http://www.zerowastescotland.org.uk/circular-economy/business-support-service.

———. 2017c. *Circular economy investment fund* [Online]. Available: http://www.zerowastescotland.org.uk/circular-economy/investment-fund.

———. 2017d. *Meeting regulations and legislation* [Online]. Available: http://www.zerowastescotland.org.uk/content/meeting-regulations-legislation.

———. 2017e. *New app makes going green easy for retailers* [Online]. Available: http://www.zerowastescotland.org.uk/content/new-app-makes-going-green-easy-retailers.

———. 2017f. *Scottish circular economy business network* [Online]. Available: http://www.zerowastescotland.org.uk/circular-economy/scottish-network.

11

Best Practices

Introduction

The following best practices have been identified from the case studies of Seattle, London, Flanders, New South Wales, Denmark, Germany, the Netherlands, and Scotland implementing a variety of fiscal and non-fiscal tools to develop the circular economy and encourage a life cycle perspective to be taken by economic actors in an attempt to decouple economic growth from resource use and associated environmental impacts. These best practices can be implemented by other locations around the world attempting to develop the circular economy.

Upstream

From the case studies, a variety of best practices have been identified in the use of upstream fiscal and non-fiscal tools to develop the circular economy and encourage a life cycle perspective to be taken by economic actors in an attempt to decouple economic growth from resource use and associated environmental impacts.

© The Author(s) 2018
R. C. Brears, *Natural Resource Management and the Circular Economy*,
Palgrave Studies in Natural Resource Management,
https://doi.org/10.1007/978-3-319-71888-0_11

Environmental Taxes and Charges

Governments use a variety of environmental taxes and charges to price-in environmental costs and allow businesses and consumers to determine how best to reduce their environmental footprint.

Tax Breaks for Companies

One location offers tax breaks for companies to develop new environment-friendly products and technologies as well as a tax break for companies in any sector to invest in new resource-efficient technologies. Similarly, one jurisdiction encourages investments in energy-efficient technology by offering tax deductions on income tax. The same jurisdiction provides a tax refund on general environment-friendly investments as well as a scheme that enables entrepreneurs to decide when to write off their environmentally friendly investment costs and how fast or slow that should be. Finally, one another location offers a tax incentive that enables individual investors to invest in green projects with the size of the incentive dependent on the size of the investment made.

Industry-Specific Tax Breaks

Tax breaks can be created to encourage resource efficiency in specific industries. For example, one location has created a Fertilizer Account for farmers with the criteria that they submit fertiliser plans. Once doing so, the farmers can buy chemical fertilisers without paying the associated tax on the fertiliser.

Subsidies and Incentives

Subsidies and incentives are used to stimulate the development of new technologies, help create markets for circular economy goods and services, encourage changes in consumer behaviour, and temporarily support higher levels of environmental protection by companies.

Funding Circular Economy Innovations

Jurisdictions provide a variety of funding options for businesses to develop circular economy innovations that promote resource efficiency and better waste recycling. One location offers tailored financial instruments including loans to circular economy business projects that use renewable inputs, recover value, and prolong product life. Another jurisdiction offers enterprises a subsidy for implementing environmental technologies including energy and renewable energy technologies that are on a limited technology list, with the size of the subsidy being dependent on the technology's performance. The same jurisdiction also offers grants for strategic projects that focus on closing the loop with the size of the grant dependent on the performance of the technology. Other locations offer funding for businesses developing innovative new circular economy business models as well as funding for the establishment of green industrial symbioses that promote economic competitiveness and resource efficiency.

Funding Large-Scale Demonstration Projects

One jurisdiction provides demonstration funding for large-scale projects that show for the first-time advanced methods for avoiding or reducing environmental pollution, with two types of funding available: an investment grant or an interest subsidy. Another location launched a fund that supported major demonstration projects that provided transformative and innovative local energy solutions, with projects receiving funding expected to demonstrate partnership and collaborative learning as well as add local value.

Supporting SMEs

To support small and medium-sized enterprises (SMEs) in developing circular economy products and services, one location offers SME loans to improve energy and material use efficiency. In addition, the location offers SMEs funding to develop waste-prevention measures. Finally, the same location offers SMEs circular economy investment funding to help develop circular economy business models.

Funding Resource Efficiency in Daily Operations

To reduce resource consumption in daily operations, one administration provides grants to SMEs to increase the numbers of energy consultations carried out and to encourage energy savings through the implementation of energy efficiency measures. Similarly, another jurisdiction provides businesses with rebates for energy efficiency upgrades with the rebate amount determined by the cost savings of the measures taken. Finally, to turn waste into a resource, one location offers low-interest financing for large-scale investments that prevent or use waste heat in business operations.

Promoting R&D

One location has an innovation fund that provides financial support to research and development (R&D) projects focusing on advances in bio-resources, energy, climate, and materials with funding available for all types of projects ranging from large-scale projects to entrepreneurial graduate projects. Similarly, another government provides R&D grants to businesses developing new products, processes, and services that improve company competitiveness and benefit the environment.

Tradeable Permits

Tradeable permits are flexible, market-based tools that can address environmental problems. Typically, tradeable permits are available through a cap-and-trade system or as a minimum performance commitment for baseline and credit schemes. One jurisdiction offers financial incentives for businesses to voluntarily invest in energy saving projects with participating businesses receiving certificates for energy saved. These certificates can then be sold on to mandatory participants including electricity retailers and direct suppliers of electricity.

Regulations

Regulatory tools represent a major proportion of all tools used for environmental policy. The appeal of regulations is that rather than dictate the specific technique for reducing pollution, regulations grant flexibility in choosing how to meet a regulation and therefore are more cost-effective than specific technology mandates.

Environmental Permits

One location requires that any company that sets up in the region must check to see whether they require an environmental permit. Whether a permit is required is based on which category the company falls into, with companies having high levels of nuisance to humans and the environment requiring a permit. Similarly, another location requires all industrial installations have an environmental permit that establishes limits for the discharge of substances that could pollute water, soil, and air.

Removing or Overcoming Regulatory Hurdles

To remove regulatory hurdles that hamper the development of the circular economy, one location established a task force for resource efficiency that reviewed existing regulations for any hurdles in addition to developing solutions to overcome specific barriers to businesses using resources and materials more efficiently. Likewise, another jurisdiction has developed a smart regulation programme that provides support to businesses that wish to invest in circular economy innovations but encounter legislative and regulatory barriers. A multidisciplinary team will interview the business and consult all relevant stakeholders to find a solution.

Green Public Procurement

Governments can use green public procurement policies to increase their credibility when encouraging industry and consumers to change their patterns of production and consumption. Green public procurement policies can also be used to enlarge the market for sustainable products and services. One location offers a range of tools including free-of-charge support to public organisations conducting energy retrofits as well as a framework for public organisations to use when procuring energy service companies to ensure they achieve energy and cost savings.

Cluster Policies

Many governments have initiated cluster policies that enable firms to benefit from industrial symbioses, for example, the exchange of by-products among companies. One location has developed a range of clusters that contribute to the region's aim of decreasing waste generated from companies with cluster participants utilising waste streams, using raw materials more efficiently, and exchanging residues between companies. One such project involves matching companies' waste streams with other companies seeking to turn waste into resources. To ensure clusters are run as efficiently as possible, one jurisdiction provides support to cluster management organisations across a variety of industries and technological sectors. Members of the programme are required to participate in a benchmarking and certification process with the associated costs fully funded by the government. In addition, the same location created a cluster competition in which clusters with the best strategies for future markets received funding over a five-year period.

Enhancing Business Competitiveness

Governments are recognising the importance of environmental management as a competitive advantage and as a driver of the circular economy. As such, governments are supporting industry sectors and businesses through the promotion of practices that enable them to trade based on sustainable and efficient business practices.

Partnerships to Showcase Innovations

One location has set up an initiative, in partnership with a business association, to showcase local entrepreneurs and start-ups that are developing circular economy products and services across a variety of economic sectors.

Online Portal for Best Practices

A government has established an online portal that encourages environmental technology transfer between companies with the portal containing publications on research findings and best practice examples from a range of cleaner production topics. To ensure credibility, all publications have been reviewed by experts.

Flagship Projects Leading by Example

To promote resource efficiency in businesses, one government has developed an energy modernisation programme, in partnership with the commercial and real estate sectors, to facilitate flagship redevelopment projects that can be imitated by other commercial properties across the country.

Education and Training

The transition towards a circular economy requires a requisite skills base of not only trained professionals but of society too. One jurisdiction selects several circular economy champions per year and provides each with education and skills development to promote the circular economy within their organisations. Similarly, another location provides training programmes for businesses to help them gain a competitive advantage through resource and energy efficiency.

Raising Industry Awareness and Capacity

Disseminating and demonstrating the benefits of circular economy practices and technologies is an effective way of fostering the adoption of new production methods.

Support for SMEs

One location provides free, practical advice to SMEs to help them adopt and scale up circular economy business models with help including the deployment of expert advisors. Similarly, another administration provides one-to-one support to small and medium-sized businesses to help them identify resource efficiency-related cost-saving opportunities.

Industry-Specific Guidelines

One authority has provided sustainable design and construction guidance for architects, developers, and engineers to achieve sustainable development in the built environment. Similarly, another jurisdiction has created guidelines for the construction industry to design out waste across the entire construction process. Meanwhile, another location has developed guidelines for the built environment that includes best practices on how to meet citywide sustainability targets. Similarly, another location has developed guidelines for professional event caterers to plan and consider all steps in reducing food wastage.

Tailored Advice

One government has launched a programme in which advisors work with businesses to identify opportunities to achieve clean energy targets including developing on-site renewable energy or off-site procurement of renewable energy. Meanwhile, another location's electricity provider works with designers to help them integrate energy efficiency into building projects from the beginning. Another authority has developed an industry-specific energy efficiency initiative that provides participants with coaching and helps in implementing energy-saving measures.

Free Tools and Assistance

One administration provides free tools and assistance to help local businesses conserve resources and prevent pollution with advice given in the areas of recycling, preventing water pollution, saving water, and how to save money from rebates and self-audits. Similarly, another authority provided a material scan for SMEs to encourage them to look at ways of reducing raw material usage with the results reviewed by specialised consultants.

Industry-Based Standards

Industry-based standards help define and promote resource efficiency standards for products and services. One location has established an industrial agreement to ensure that only sustainable biomass is used in combined heat and power stations, with annual reporting either developed or verified by a third party and made publicly available. Meanwhile, another location has set up a nutrient platform that involves stakeholders from a variety of sectors along with government ministries joining forces to close nutrient cycles and create a market for recycled nutrients.

Voluntary Agreements

Voluntary agreements encourage businesses, industries, and sectors to improve their resource efficiency and environmental performance beyond regulatory measures. One jurisdiction has created an energy agreement between the government, employers, trade unions, and environmental organisations to achieve energy conservation, increase renewable energy produced, and create jobs. The same location has developed an agreement on creating a sustainable garment and textile sector with the government, industry organisations, trade unions, and civil society organisations seeking to reduce the environmental impact of raw materials production as well as use fewer resources in the industry.

Supporting Life Cycle Analysis

A life cycle analysis seeks to minimise environmental impacts by examining all phases of a product's or service's life cycle and acting where it is most effective.

Life Cycle Analysis Tools for Products and Services

One administration has developed an eco-design tool that helps designers and companies calculate quickly and easily the environmental impact a product has throughout its life cycle. The same administration has also created a toolkit that helps employees visualise how new sustainability opportunities can be integrated into the innovation and design process of both products and services.

Industry-Specific Life Cycle Tools

One government has created a free, online life cycle assessment database that contains quality-checked data sets from the construction sector for all relevant building materials, enabling users to determine a building's ecological quality. This is in addition to a life cycle analysis tool for buildings that contains construction material data sets in a preconfigured way for users to conduct uniformed life cycle assessments across a variety of buildings.

Environmental Recognition Awards

Environmental recognition awards help raise environmental awareness of businesses, the community, and individuals and help companies gain recognition for their environmental performance.

Awards for Businesses

To encourage businesses to reduce resource consumption, one jurisdiction has initiated a business energy challenge in which businesses are awarded for their efforts in reducing energy use (and carbon emissions) across their property or property portfolio. This is in addition to a clean city awards scheme in which participants are recognised and rewarded for circular economy good practices. Another administration runs a sustainability advantage programme in which participants receive expert advice, training, guides, case studies, one-on-one support as well as networking opportunities, with organisations publicly recognised for real progress made. Similarly, another location has an eco-design award for companies of all sizes, across all sectors, that have developed a product, service, or concept that demonstrates a high level of innovation from both a design and environmental perspective, with applications assessed on the entire life cycle of the product, service, or concept.

Awards for Younger Generations

One location has developed an entrepreneurs' award to encourage the development of the circular economy within younger generations with students submitting their ideas on how to use resources more efficiently with the winner receiving a cash prize. Meanwhile, another location runs an eco-efficiency award to raise awareness among students on the circular economy.

Knowledge Transfer Networks

Governments can facilitate circular economy linkages between enterprises, institutions, universities, private research labs, SMEs, consultants, and entrepreneurs through knowledge networks. A jurisdiction has created a public-private partnership between the city, government agencies, business, and academia to develop alternative fuels with the initiative hosting dialogues, offering funding opportunities, and holding forums to share knowledge. Meanwhile, another location has developed a circular economy business network that encourages innovative, forward-thinking businesses to work collaboratively and share best practices to help develop the circular economy.

Information-Based Tools

Good quality information is essential for identifying environmental challenges and informing consumption and production decisions. One administration has developed an online eco-efficiency scan that enables organisations to quickly identify measures that can increase eco-efficiency of any business. The scan also directs users to where they can get specialised advice to increase their efficiency. Likewise, another location has developed an online energy efficiency portal so that a variety of commercial and administrative buildings can find information about energy savings and promotion programmes as well as search for qualified energy consultants and experts.

Downstream

From the case studies, a variety of best practices have been identified in the use of downstream fiscal and non-fiscal tools to develop the circular economy and encourage a life cycle perspective to be taken by economic actors in an attempt to decouple economic growth from resource use and associated environmental impacts.

Environmental Taxes and Charges

Governments use a variety of environmental taxes and charges to price environmental costs and allow businesses and consumers to determine how best to reduce their environmental footprint.

Deposit Schemes

Locations have implemented deposit schemes for one-way drinks with one place installing reverse vending machines in shops and supermarkets to enable consumers to deposit their containers in exchange for a refund slip that can be cashed in at the store. Meanwhile, another location operates a deposit-return scheme for both one-way packaging and refillable bottles with the amount refunded to customers depending on the type of material used in the bottles and cans, the volume of each bottle or can, and whether the bottle or can will be recycled or reused.

Packaging Tax

One administration has implemented a packaging tax to reduce waste and increase the reuse and recycling rate of packaging. The tax is divided into a volume-based tax and a weight-based tax. Another jurisdiction requires that all producers and importers creating a certain amount of packaging are required to pay a fee per kilogram of material used. Finally, another authority has established a carrier bag charge for single-use carrier bags.

Subsidies and Incentives

Subsidies and incentives are used to stimulate the development of new technologies, to help create markets for circular economy goods and services, encourage changes in consumer behaviour, and temporarily support higher levels of environmental protection by companies.

Funding Circular Economy Projects

One administration provides funding for businesses to carry out waste assessments, install on-site small-scale recycling equipment, and work with experts to identify projects that will reduce waste or provide resources to create new material. Another location provides funding for the conducting of small, targeted projects related to the circular economy with a specific focus on projects that prevent waste and encourage high-quality reuse of materials. Finally, one location will distribute funding over the next decade to improve air quality across the city with funding available to projects that reduce pollution at construction sites.

Developing New Recycling Infrastructure

One jurisdiction is providing funding opportunities for organisations to develop projects that will create new recycling infrastructure solutions, establish or expand recycling material markets, and increase the efficiency of recycling facilities for specific targeted wastes. In addition, funding is available for resource recovery facility expansions and enhancement projects that increase recycling of household, industry, and business waste.

Regulations

Regulatory tools represent a major proportion of all tools used for environmental policy. The appeal of regulations is that rather than dictate the specific technique for reducing pollution, regulations grant flexibility in choosing how to meet a regulation and therefore are more cost-effective than specific technology mandates.

Mandatory Reporting

A jurisdiction requires all buildings of a certain size and over to benchmark their energy usage with the city making the data publicly available, enabling businesses and consumers to make informed decisions when buying or renting properties. Meanwhile, an administration requires all large companies to report their corporate social responsibility (CSR) policies including standards, guidelines, or principles, how they translate their policies into actions, and an evaluation of what has been achieved during the last financial year.

Mandatory Recycling

One location has initiated a range of mandatory recycling policies including prohibiting any residents or businesses from putting food scraps and recyclable materials in their rubbish with businesses and property managers required to provide convenient food and recyclable-related services at their properties. The same location requires all food service businesses to find recyclable or compostable packaging alternatives to all disposable food service items. Another administration requires all businesses and organisations to recycle a range of materials including plastics, paper, and glass in addition to requiring all food businesses which produce a certain amount of food waste to present it for separate collection.

Green Public Procurement

One jurisdiction is aiming for 100 percent sustainable procurement by ensuring the process includes life cycle thinking as well as support for public-private partnerships that encourage innovative circular economy approaches. Meanwhile, another administration is using green public procurement to show leadership in resource productivity with government agencies required to set resource consumption targets and publish annual performance reports. Finally, one authority has developed a green procurement initiative that involves public organisations at the national and regional level committing to reducing their environmental impact from procurement actions and driving the market in a greener direction.

Enhancing Business Competitiveness

One government has created a programme that involves companies and research and educational institutions collaborating on circular economy waste-to-resource designs, technologies, products, processes, and systems. Meanwhile, another location has developed a range of environmental KPIs to help businesses in a variety of sectors measure and monitor their resource usage.

Education and Training

In the transition towards a circular economy, education and training for resource efficiency should be a continuous task across the educational curriculum that endorses lifelong learning.

Teaching the Circular Economy

One location has developed a toolkit for schools that guides teachers on how to integrate the concept of the circular economy into education programmes with the kit including a guide as well as informational cards on various aspects of the circular economy including life cycle thinking and eco-design.

Circular Economy Training Courses

For professionals, one administration has developed half-day courses in which participants will come away with knowledge on what circular economy business strategies are available and how the circular economy can provide numerous ecological and business opportunities.

Raising Industry Awareness and Capacity

Disseminating and demonstrating the benefits of environmentally sustainable and circular economy practices and technologies is an effective way of fostering the adoption of new production methods. Governments have created a wide array of tools to help businesses become more circular.

Audits and Reviews

One location provided advice to businesses on how they could reduce their food waste and put surplus food to good use with an organisation providing food waste audits for participating businesses. Another location has developed a green event scan tool that helps event organisers reduce material and resource use. The tool provides tips on how to make improvements along with a total score that can be shared on social media.

Guidance on Avoiding Waste

Locations have developed best practice guidelines for commercial and industrial facilities to avoid waste and increase the yield and quality of recyclable materials. Meanwhile, one specific jurisdiction has developed guidelines for businesses to reduce waste that includes case studies as well as practical information on how to avoid waste and increase recycling rates. Similarly, another location has developed guidelines to help businesses measure and manage their resource consumption.

Industry-Based Standards

One location has developed an agreement between the private sector and the government to start green projects together with all parties committing themselves to realising projects within specific timeframes. Meanwhile, another location has a packaging covenant that involves the public and private sector funding projects that address sustainable packaging-related issues. Finally, another location has developed an initiative, involving government and civil society, that aims to help businesses translate CSR opportunities directly into market opportunities.

Voluntary Agreements

One jurisdiction has created an initiative that commits participants to increase the level of circular purchases being made. Over a two-year period, participants will create circular acquisition projects and develop, share, and disseminate knowledge about circular purchasing. Another administration has developed a packaging agreement that sets out a series of national recycling targets for packaging materials and requires relevant industry participants to develop packaging waste-prevention plans. Finally, one location has developed agreements in which building owners, finance providers, and local governments come together to implement resource efficiency upgrades in buildings with finance providers making finance available for building owners and the local government levying a charge for the environmental upgrades, with the collected levy passed onto finance providers to repay the funds advanced.

Eco-Labels and Certification

In one jurisdiction, energy labelling is mandated when selling or letting buildings so buyers and renters can make informed decisions as well as determine which energy-related retrofits can be implemented cost-effectively. One government runs an eco-label scheme that guarantees labelled products or services meet high standards in terms of environmental protection, health, and performance across its entire life cycle. To ensure the labelling process is up-to-date, the criteria used for assessment is reviewed every several years. In addition, the same government offers a car label that informs consumers on how efficient specific vehicle models are. Regarding reusing of products, one location offers a reuse quality standard for shops that sell second-hand goods with the quality standard being a sign of quality, reliability, and professionalism.

Supporting Life Cycle Analyses

To help government agencies procure energy-using products and services, one government agency has developed Excel models that enable public-sector procurers to calculate procurement, operating, and disposal costs as well as the life cycle costs of a range of specific products.

Greening the Supply Chain

Companies are often encouraged or mandated by governments to ensure environmental standards are maintained throughout the entire supply chain. In this way, buyer companies can ensure that the environmental standards that they have adopted internally are being consistently maintained by their suppliers.

Public-Private Information Sharing

One jurisdiction has created a forum on sustainable procurement that increases the awareness of professional buyers, from both public and private sectors, on the benefits of sustainable procurement. Similarly, another government has created an initiative that supports entrepreneurs and the government in making circular purchases where products or materials can be used in a new cycle. The initiative will involve meetings to develop new ideas, solve problems, and achieve various innovations with working groups established across a variety of sectors.

Guidance on Sustainable Procurement

One authority has created an Internet web page in which procurers can find environmental clauses that are ready to be inserted into tender documents for a variety of products. Meanwhile, another location has created a guideline for SMEs on sustainable procurement and how they can influence their customers in a positive way about the goods and services they buy and use.

Environmental Recognition Awards

One authority has created a competition in which people pitch projects that aim to develop the local circular economy, with the winner receiving a cash prize along with cash raised through a crowdsourcing platform. The same authority also held a hackathon, open to entrepreneurs, start-ups, as well as small businesses, to showcase innovative ideas that deliver solutions for improving the circularity of the economy with the winning entry receiving a cash prize. Another location offers an award for green events with a cash prize for small, medium, and major-sized events. Finally, one government has created a resource efficiency pledge for businesses with participating businesses receiving recognition awards for using resources more efficiently.

Extended Producer Responsibility

One jurisdiction requires producers, as well as importers, exporters, and distributors, to assume responsibilities for the entire life cycle of electronic and electric products with local collection points established for producers and so on to collect disposed items from. The same location also requires retail stores of a certain size and above to take back, upon sale of a new product, a used device of the equivalent type free of charge.

Knowledge Transfer Networks

One administration has developed a cooperative network between the government and stakeholders from the construction industry to promote the sustainable management of materials including ensuring waste is of a high quality for recycling. Similarly, another location has developed a construction industry partnership network, involving public and private parties, to encourage the industry to undertake best practice pollution abatement measures on construction sites. The same location has also established an alternative fuel network to encourage the fuel industry and local authorities to develop biodiesel for public transportation. Meanwhile, another government has developed a recycling technologies network to bring together companies and government institutions across the whole sector to strengthen the competitiveness of SMEs and promote the marketing of innovative recycling technology.

Information-Based Tools

Good quality information is essential for identifying environmental challenges and informing consumption and production decisions.

General Resource Mapping

One location has produced an online map that publicly recognises businesses that have acted to reduce resource consumption, waste, and pollution. To be put on the map, a business must take a minimum amount of actions; with more actions taken, the darker a business's icon appears on the map. Another administration provides a natural capital atlas that enables businesses to see how natural resources can be used more sustainably and efficiently, with the information able to be incorporated into CSR reports.

Resource-Specific Mapping

One jurisdiction has created an interactive online platform that lets users explore the region's renewable energy potential using geographic information systems software. Another government has created an online heat map that allows users from both public and private sectors to find opportunities for decentralised energy projects with the map providing data on major energy consumers, community heating networks, and heat density. Similarly, one jurisdiction has created a waste map that visually presents waste data with the map identifying operational waste sites as well as potential waste sites for waste disposal, organic treatment, and material recycling.

Spreadsheet Tracking

To help businesses monitor their resource use, a jurisdiction provides businesses with a water usage tracking spreadsheet template as well as a template to measure and monitor energy usage. Similarly, another location has developed a raw material and waste tracking spreadsheet to show the true cost of waste and help focus efforts on areas where the biggest cost savings can be made.

Planning Tools

One administration has created a waste assessment tool to help businesses reduce their waste and increase recycling. The tool involves a four-step process in which businesses plan to conduct a waste audit, gather data, analyse the results, and act. Another location has created an industry-specific waste management planning tool that helps companies track materials and waste on construction sites. Regarding recycling, one administration has developed an interactive online tool that calculates the environmental benefits of large-scale recycling initiatives across a variety of material types with stakeholders able to estimate the full environmental benefits of recycling.

Circular Economy Purchasing

An administration has created a clean vehicle checker tool that helps consumers buy less polluting vehicles, with the tool enabling users to find out the emissions of specific vehicle types and models. Meanwhile, another location has developed a smartphone app that provides retailers with tools and information to use resources as efficiently as possible. The app also provides a link to free, tailored advice and support.

12

Conclusions

In our current economic model, the amount of waste has constantly grown as economic growth has been based on a 'take-make-consume-dispose' model. This linear model assumes that resources are abundant, available, and cheap to dispose of. While the current model has generated an unprecedented level of growth, the model has led to constraints on the availability of natural resources due to rising demand, generation of waste, and environmental degradation associated with a variety of challenges.

A business-as-usual approach towards economic growth will result in the need for the equivalent of two planets to sustain us. In the linear economy, as income levels rise, demand for various products including foods and electronics will increase. At the same time, raw materials are becoming critically short in addition to becoming poorer in quality due to over-exploitation of higher grade ores. Resource prices will become more volatile as the century progresses which could jeopardise peace and security, particularly in low-income countries. Rapid population growth and urbanisation will see an increase in demand for energy as well as water resources. In addition, population growth will result in an increased demand for new infrastructure, including clean technologies and water

© The Author(s) 2018
R. C. Brears, *Natural Resource Management and the Circular Economy*,
Palgrave Studies in Natural Resource Management,
https://doi.org/10.1007/978-3-319-71888-0_12

resources. There will also be significant costs of adapting new infrastructure to climate change-related risks. With economic growth, the world will see an increase in demand for energy with fossil fuels likely to be the dominant source of energy for years to come. The world's water resources will likely be over-exploited due to increased demand from industry and agriculture, with many related activities diminishing the quality of water available to other users too. With increased urbanisation, the amount of waste generated will rise as will air pollution with greater numbers of people being exposed to higher levels of pollution. The combined trends of rising economic and population growth will further deteriorate ecosystem services that are vital to human health and well-being. Finally, climate change-related extreme weather events will have a significant impact on economies around the world if emission levels continue a business-as-usual trend.

In contrast, the circular economy focuses on recycling, reducing, and reusing physical resources and using waste as a resource with the overall goal of decoupling economic growth from resource use and associated environmental impacts. An important aspect of the circular economy is recognising product life cycles in which a product's life can be divided into stages that transition from upstream to downstream. This means that in the circular economy, products are designed so that waste is phased out in both design and production, rather than relying on waste disposal solutions at the end of a product's life.

Government intervention has an important role in developing the circular economy and encouraging a life cycle perspective to be taken by economic actors. Governments can use a variety of innovative fiscal and non-fiscal tools to encourage economic actors to design out waste throughout the value chain, facilitate access to financial capital for businesses developing circular economy innovations, provide research and development (R&D) funding for circular economy technologies, support entrepreneurs and small-to-medium enterprises developing new circular economy products, as well as facilitate better consumption choices of consumers. Fiscal tools help economic actors consider the hidden cost of production and consumption. Environmental taxes are commonly applied to energy and water usage while charges are levied on economic actors who use the environment. The aim of environmental taxes and charges is to prevent or reduce potential harm relating to natural resources use and encourage proper waste disposal. Subsidies and incentives are used to encourage the development of new circular economy technologies and help create markets for recovered raw materials as well as change consumption choices made by consumers. Tradeable permits, in the form of cap and trade schemes and baseline and credit schemes, encourage participants to develop innovative solutions to lower pollution limits and ensure minimum environmental outcomes are achieved.

Non-fiscal tools are used to encourage reductions in resource consumption and the uptake of circular economy technologies. Rather than set specific technology standards to enhance resource efficiency and lower pollution levels, regulations are often used to set environmental performance standards, allowing flexibility in how those affected can meet the standard in a cost-effective way. Green public procurement policies are used to show government leadership as well as enlarge the market for circular economy-related products and services. Governments use a variety of tools to simultaneously enhance the economic and environmental performance of businesses with examples being the facilitating of knowledge networking opportunities on circular economy goods and services and providing of market intelligence on the clean technology sector. Cluster policies are developed to enable firms to benefit from industrial symbioses where by-products are exchanged among companies to create new materials. Education and training opportunities are provided to enhance awareness of the various environmental and economic benefits circular economies provide with resource efficiency-related education provided in schools, technical and vocation programmes, higher education, and on the job. Governments often develop programmes to raise industry awareness and capacity on the benefits of environmental sustainability and resource-efficient practices and technologies with tools commonly used including resource efficiency saving advice and funding opportunities for businesses to develop circular economy products. Industry-based standards are developed by firms to define and promote standards, enabling all participants to comply with environmental standards that are set by governments. Voluntary agreements involve the government working with businesses, industries, or sectors to improve their resource efficiency beyond regulatory measures with multiple benefits including increased buy-in from businesses and a more proactive mindset towards developing circular economy innovations. Eco-labels and certification provide consumers with the environmental impact of products and can be applied to the entire product life cycle or a specific step or stage in the production process. Governments often provide procurers with tools to assess the life cycle costs of various products. To encourage the greening of supply chains, either voluntary or mandatory measures are developed to ensure environmental standards are maintained throughout the entire supply chain. Environmental recognition awards help individuals, businesses, and communities gain recognition for exceptional environmental performance. Extended producer responsibility initiatives involve take-back programmes and disposal restrictions to ensure producers take responsibility for products beyond the post-consumer stage. Knowledge transfer networks are frequently developed by governments to help facilitate circular economy partnerships between companies, research institutions, and entrepreneurs. Finally, information-based tools are developed to help economic actors identify challenges and make informed consumption and production decisions.

From the case studies of London, Seattle, Flanders, New South Wales, Denmark, Germany, the Netherlands, and Scotland using fiscal and non-fiscal tools, both upstream and downstream, to develop the circular economy, the following best practices have been identified for other locations around the world attempting to develop the circular economy.

Upstream, tax breaks can be used to encourage companies to develop new circular economy products as well as invest in resource-efficient technologies. Subsidies and incentives are used to stimulate the development of new technologies as well as develop markets for circular economy products; for example, governments can offer tailored loans for circular economy business projects as well as subsidies for environmental technologies with the size of the subsidy based on the technology's performance. Funding is often made available for large-scale demonstration projects that showcase for the first-time advanced methods for avoiding or reducing resource use. At the firm level, grant funding is frequently provided for small and medium-sized enterprises to improve resource efficiency. Finally, funding is made available for R&D projects that focus on new circular economy-related products, processes, and services that enhance economic and environmental performance. Tradeable permits are used to reduce demand for resources and lower carbon emissions, with one location providing energy saving certificates for businesses that can be sold on. Regulations, including environmental permits, are often used to encourage companies to reduce their environmental footprints with governments establishing pollution limits. In addition, regulatory task forces can be established to ensure regulations do not unnecessarily impact investments in circular economy technologies. Green public procurement policies help government agencies increase their credibility when encouraging economic actors to be more sustainable with policy tools used including resource retrofits as well as frameworks to follow when procuring resource efficiency upgrade services. The productivity of circular economy clusters can be enhanced with support provided to better match companies' waste streams as well as enhance the efficiency of management practices. Business competitiveness can be boosted with support provided to companies to showcase their technologies to investors. In addition, online portals can be developed that enable companies to find best practices to emulate. Education on the circular economy can include business education and skills development programmes to promote circular economy thinking in the product design stage. Raising industry awareness and capacity on circular economy issues can be facilitated by providing support to businesses in adopting and scaling-up circular economy business models, developing industry-specific guidelines to design out waste, and offering tailored circular economy advice including free tools and assistance to help businesses reduce resource usage and waste. Industry-based standards can help sectors reduce material usage, environmental degradation, and carbon emissions, for instance, encouraging the use of sustainable biomass in energy production. Meanwhile, voluntary agreements can be established between government agencies and industry to establish resource use targets in the delivery of goods and services including energy reduction targets and better environmental outcomes in the textile and garment sector. Governments can support life cycle analysis concepts by providing tools that help designers calculate the environmental impact of products throughout their life cycle as well as industry-specific tools that help determine resource use of buildings. Environmental recognition awards can be used to encourage businesses to reduce resource consumption with support provided to participants including expert advice, training, case studies, and networking opportunities. Awards can also be created for students to encourage the development of the circular economy within younger generations. Knowledge transfer networks can be created to facilitate circular economy linkages between the public, private, and research sectors on issues including alternative fuel development. General circular economy business networks can be established to facilitate exchanges of ideas and best practices. Information-based tools can be developed to identify environmental challenges and make informed consumption decisions, for instance, eco-efficiency scans to help organisations easily and quickly identify measures to enhance resource efficiency. Online portals to enhance resource consumption decisions can also be developed such as energy efficiency portals that provide energy saving tips as well as a database to search for energy consultants.

Downstream, deposit schemes encourage consumers to return one-way prescribed beverage containers for a refund while other locations offer deposit-return schemes for both one-way packaging and refillable bottles. Packaging taxes can also be implemented to encourage producers and importers to reduce the amount of packaging. Subsidies and incentives can be developed to encourage businesses to conduct waste assessments, initiate resource efficiency-related projects including ones that reuse materials, and develop recycling infrastructure facilities for specific targeted wastes. Regulations can be developed to ensure mandatory reporting of buildings' and companies' resource usage as well as corporate social responsibility (CSR) policies. Mandatory recycling policies can be implemented to reduce both residential and commercial recyclable materials from going to landfills. Green public procurement policies can set resource reduction targets as well as ensure environmental factors are considered when making procurement decisions on products. To help enhance business competitiveness, governments can create waste-to-resource programmes that bring together companies, research institutions, and so forth to develop circular economy designs and technologies. Key performance indicators help businesses across a variety of sectors reduce their energy and water usage and waste. Educational and training programmes can integrate circular economy concepts in school education programmes to encourage future consumers to reduce their environmental footprints. Programmes can be developed for employees to understand what is the circular economy and strategies available to reduce resource demand. To raise industry awareness and capacity, resource audits can be provided to help businesses reduce their waste as well as tools that help businesses determine where resource savings can be made. Best practice guidelines raise awareness on how to reduce waste and increase the quality of recyclable materials in a variety of sectors including construction and demolition. Industry-based standards can involve public and private sector actors partnering on initiatives that reduce packaging as well as initiatives that translate CSR into business opportunities. Voluntary agreements can be developed for circular economy purchasing, packaging agreements, and even building upgrades. Eco-labels and certifications can be applied to specific products, such as buildings, so buyers and renters can make informed decisions, or to entire product ranges; for example, an eco-label for any goods or services that meet high environmental protection and performance standards across the entire life cycle. Life cycle analysis tools can be created so purchasers from government agencies can calculate the entire life cycle costs of a range of products. To promote the greening of supply chains, public sector-initiated forums can raise awareness of procurers from both public and private sectors on the benefits of circular economy purchasing. Guidelines can also be provided to encourage procurers from the private sector to factor in resource efficiency in their decisions. Environmental recognition awards can be developed for individuals, entrepreneurs, and companies proposing new solutions that encourage consumers to reduce their environmental footprints. Pledges encourage businesses to make public pledges to use resources more efficiently in day-to-day operations. Extended producer responsibility schemes can mandate that retailers collect from consumers used electronic and electric waste products. Knowledge transfer networks can promote the abatement of pollution in a variety of industries including the construction industry as well as bring companies together to share knowledge and expertise on recycling technologies. A variety of information-based tools can be developed to encourage lower environmental footprints, for instance, green business maps that publicly display businesses making commitments to reducing resource use, pollution, and waste. Resource mapping tools can show the potential for resource recovery in specific areas, with examples including a heat map that determines the potential for using waste heat as a resource and a map that shows potential sites for material recycling. Finally, spreadsheet templates can be developed for businesses to measure and monitor their resource consumption levels including energy and water as well as track their waste. Planning tools can be developed to help businesses reduce their waste and increase recycling with tools able to estimate the full environmental benefits of recycling. To encourage circular economy purchasing, online tools can be created to help consumers decide between products; for example, one tool enables users to compare the emissions of various car makes and models. Smartphone apps can also be developed for retailers to use resources more efficiently with access to free, tailored advice and support.

In conclusion, developing the circular economy is not a static activity. Instead, it requires a variety of fiscal and non-fiscal tools to develop the circular economy and encourage a life cycle perspective to be taken by economic actors in an attempt to decouple economic growth from resource use and associated environmental impacts.

Index

© The Author(s) 2018
R. C. Brears, *Natural Resource Management and the Circular Economy*,
Palgrave Studies in Natural Resource Management,
https://doi.org/10.1007/978-3-319-71888-0

Printed by Printforce, the Netherlands